CPC UCW
LLYFRGELL
LIBRARY

Nonlinear Problems in Abstract Cones

This is Volume 5 in
NOTES AND REPORTS IN MATHEMATICS IN SCIENCE AND ENGINEERING

Edited by William F. Ames, *Georgia Institute of Technology*

A list of recent titles in this series appears at the end of this volume.

Nonlinear Problems in Abstract Cones

Dajun Guo
Shandong University
Jinan
People's Republic of China

V. Lakshmikantham
University of Texas at Arlington
Arlington, Texas

ACADEMIC PRESS, INC.
Harcourt Brace Jovanovich, Publishers
Boston San Diego New York
Berkeley London Sydney
Tokyo Toronto

ACADEMIC PRESS, INC.
1250 Sixth Avenue, San Diego, CA 92101

United Kingdom Edition published by
ACADEMIC PRESS INC. (LONDON) LTD.
24–28 Oval Road, London NW1 7DX

Library of Congress Cataloging in Publication Data

Guo, Dajun.
Nonlinear problems in abstract cones/ Dajun Guo.
V. Lakshmikantham.
p. cm. – (Notes and reports in mathematics in science and engineering : v. 5)
Bibliography: p.
ISBN 0-12-293475-X
1. Cones (Operator theory) 2. Fixed point theory. 3. Integral equations, Nonlinear. 4. Differential equations, Nonlinear.
I. Lakshmikantham, V. II. Series.
QA329.2.G86 1988
515.7'24 – dc19 87-27358
CIP

88 89 90 91 9 8 7 6 5 4 3 2 1
Printed in the United States of America

TABLE OF CONTENTS

PREFACE

In many problems that arise from models of chemical reactors, neutron transport, population biology, infectious diseases, economics, and other systems, we need to discuss the existence of nonnegative solutions with certain desired qualitative properties. What we normally understand by nonnegativity can be developed by arbitrary cones, that is, closed convex subsets of the space under consideration. These cones define a relation "$\leq$" by means of which certain elements can be compared better than crude estimates in terms of a norm. One way to extend the notion of monotonicity to higher dimensions is to utilize the relation induced by a cone. Furthermore, the laws of nature prescribe definite bounds for solutions of real-world problems that imply the existence of solutions in conical segments. Moreover, in the study of large-scale dynamical systems by the method of vector Lyapunov functions, it is known that employing arbitrary cones offers better results than using componentwise inequalities. Thus it is clear that investigating nonlinear problems through abstract cones is an important branch of nonlinear analysis, and this book is devoted to a systematic study of this aspect, namely, the investigation of nonlinear problems in abstract cones.

Chapter 1 introduces the basic properties of cones and therefore forms a basis for the entire book. In Chapter 2, we employ the theory of cones coupled with the fixed point index to investigate positive fixed points of several classes of nonlinear operators. In Chapter 3, we utilize the fixed point theory developed to discuss positive solutions of nonlinear integral equations. Chapter 4 is devoted to the development of the theory of nonlinear differential equations in cones. Some important models that have physical meaning are worked out, and several examples from integral and differential equations are given to illustrate the abstract results. The appendix provides background material for the convenience of the reader. The notes at the end of each chapter indicate the sources of the ideas developed. References are included for further study. This work is an outgrowth of lecture notes that were used in a graduate course at Shandong University, P. R. China, and the University of Texas at Arlington. We are immensely thankful to Mrs. Sandra Weber for her excellent typing of the manuscript.

CHAPTER I. BASIC PROPERTIES OF CONES

1.0 *Introduction.*

This chapter introduces the basic properties of cones, and therefore forms a basis for the remaining chapters.

Section 1.1 deals with the normality of cones, which is the most important and useful concept. We establish some necessary and sufficient conditions for a cone to be normal.

Sections 1.2 and 1.3 consider other concepts of cones. Definitions of regularity, full regularity, minihedrality, and strong minihedrality of cones are given, and the relationships among these concepts are discussed. We also give some examples of cones that possess some or all of these properties.

In Section 1.4, we introduce the dual cone of a given cone and discuss how to use it to investigate the original cone. In addition, we give an extension theorem of positive bounded linear functionals.

Finally, Section 1.5 is divided into two parts. The first part deals with e-norm, the role of which is to produce a new solid cone to be used in some problems of differential equations. The second part is devoted to the so-called Hilbert's projective metric, which is convenient for some classes of nonlinear operators.

1.1 *Normal Cones.*

Definition 1.1.1. Let E be a real Banach space. A nonempty convex closed set $P \subset E$ is called a cone if it satisfies the following two conditions:

(i) $x \in P$, $\lambda \geq 0$ implies $\lambda x \in P$;

(ii) $x \in P$, $-x \in P$ implies $x = \theta$, where θ denotes the zero element E.

A cone is called <u>solid</u> if it contains interior points, i.e., $\overset{o}{P} \neq \phi$. A cone P is said to be <u>generating</u> if $E = P - P$, i.e., every element $x \in E$ can be represented in the form $x = u - v$, where $u, v \in P$.

Every cone P in E defines a <u>partial ordering</u> in E given by

$$x \leq y \quad \text{iff} \quad y - x \in P . \tag{1.1.1}$$

If $x \leq y$ and $x \neq y$, we write $x < y$; if cone P is solid and $y - x \in \overset{o}{P}$, we write $x << y$.

<u>Definition 1.1.2</u>. A cone P E is said to be <u>normal</u> if there exists a positive constant δ such that $||x + y|| \geq \delta$, $\forall$ $x, y \in P$, $||x|| = 1$, $||y|| = 1$. Geometrically, normality means that the angle between two positive unit vectors has to be bounded away from π. In other words, a normal cone cannot be too large.

<u>Theorem 1.1.1</u>. Let P be a cone in E. Then the following assertions are equivalent:

(i) P is normal;

(ii) there exists a constant $\gamma > 0$ such that
$||x + y|| \geq \gamma \max\{||x||, ||y||\}$ for all $x, y \in P$;

(iii) there exists a constant $N > 0$ such that $\theta \leq x \leq y$ implies $||x|| \leq N||y||$, i.e., the norm $||\cdot||$ is <u>semi-monotone</u>;

(iv) there exists an equivalent norm $||\cdot||_1$ on E such that $\theta \leq x \leq y$ implies $||x||_1 \leq ||y||_1$, i.e., norm $||\cdot||_1$ is <u>monotone</u>;

(v) $x_n \leq z_n \leq y_n$ $(n = 1,2,3,\ldots)$ and $||x_n - x|| \to 0$, $||y_n - x|| \to 0$ imply $||z_n - x|| \to 0$;

(vi) set $(B + P) \cap (B - P)$ is bounded, where $B = \{x \in E \mid ||x|| \leq 1\}$;

(vii) every order interval $[x,y] = \{z \in E \mid x \leq z \leq y\}$ is bounded.

Proof. (i)$\Rightarrow$(ii): We can assume $\|x\| = 1$ and $\|y\| \leq 1$, hence

$$1 = \|x\| \leq \|x + y\| + \|-y\| = \|x + y\| + \|y\| .$$

On the other hand,

$$\|x + y\| = \left\| x + \frac{y}{\|y\|} - \frac{1 - \|y\|}{\|y\|} y \right\| \geq \left\| x + \frac{y}{\|y\|} \right\| - 1 + \|y\| \geq \delta - 1 + \|y\| .$$

It follows therefore $\|x + y\| \geq \gamma$, where $\gamma = \delta/2$.

(ii)$\Rightarrow$(iii): Suppose (iii) is not true. Then there exists x_n, $y_n \in P$ such that $\theta \leq x_n \leq y_n$ and $\|x_n\| > n\|y_n\|$ $(n = 1,2,3,\ldots)$. Put

$$u_n = \frac{x_n}{\|x_n\|} + \frac{y_n}{n\|y_n\|}, \quad v_n = -\frac{x_n}{\|x_n\|} + \frac{y_n}{n\|y_n\|} \quad (n = 1,2,3,\ldots) .$$

It is easy to see that u_n, $v_n \in P$ and

$$\|u_n\| \geq 1 - \frac{1}{n}, \quad \|v_n\| \geq 1 - \frac{1}{n},$$

hence, by virtue of (ii),

$$\frac{2}{n} = \|u_n + v_n\| \geq \gamma\left(1 - \frac{1}{n}\right), \quad (n = 1,2,3,\ldots),$$

which is evidently impossible.

(iii)$\Rightarrow$(iv): Put

$$\|x\|_1 = \inf_{u \leq x} \|u\| + \inf_{v \geq x} \|v\| . \tag{1.1.2}$$

We prove $\|\cdot\|_1$ is a monotone norm on E. In fact, $\|\theta\|_1 = 0$ is clear. Suppose $\|x\|_1 = 0$. For any $\varepsilon > 0$ there exist $u,v \in E$ such that $u \leq x \leq v$ and $\|u\| < \varepsilon$, $\|v\| < \varepsilon$; hence, by (iii), $\|x\| \leq \|x - u\| + \|u\| \leq N\|v - u\| + \|u\| < (2N + 1)\varepsilon$. Since $\varepsilon > 0$ is arbitrary, we get $x = \theta$. The equality $\|\lambda x\|_1 = |\lambda| \cdot \|x\|_1$ obviously holds for any $x \in E$ and $\lambda \in R^1$. Now, suppose $x,y \in E$. For any $\varepsilon > 0$, there exist u_1, v_1, u_2, $v_2 \in E$ such that $u_1 \leq x \leq v_1$, $u_2 \leq y \leq v_2$ and $\|u_1\| + \|v_1\| < \|x\|_1 + \varepsilon$,

$\|u_2\| + \|v_2\| < \|y\|_1 + \varepsilon$. From $u_1 + u_2 \leq x + y \leq v_1 + v_2$ we know $\|u_1 + u_2\| + \|v_1 + v_2\| \geq \|x + y\|_1$, and so

$$\|x + y\|_1 \leq \|u_1\| + \|u_2\| + \|v_1\| + \|v_2\| < \|x\|_1 + \|y\|_1 + 2\varepsilon,$$

which, since $\varepsilon > 0$ is arbitrary, implies that $\|x + y\|_1 \leq \|x\|_1 + \|y\|_1$.

It is easy to see that $\|\cdot\|_1$ is monotone, since $\theta \leq x \leq y$ implies $\inf_{u \leq x} \|u\| = \inf_{u \leq y} \|u\| = 0$ and therefore

$$\|x\|_1 = \inf_{v \geq x} \|v\| \leq \inf_{v \geq y} \|v\| = \|y\|_1 .$$

Finally, we prove that $\|\cdot\|$ and $\|\cdot\|_1$ are equivalent. Obviously, $\|x\|_1 \leq 2\|x\|$. On the other hand, for any $u \leq x \leq v$, we have $\|x\| \leq \|x - u\| + \|u\| \leq N\|v - u\| + \|u\| \leq (N + 1)(\|u\| + \|v\|)$, hence $\|x\| \leq (N + 1)\|x\|_1$.

(iv)⇒(v): From $\theta \leq z_n - x_n \leq y_n - x_n$ we find $\|z_n - x_n\|_1 \leq \|y_n - x_n\|_1$. Since $m\|x\| \leq \|x\|_1 \leq M\|x\|$ for any $x \in E$, where $M > m > 0$ are two constants, it follows

$$\|z_n - x_n\| \leq \frac{1}{m} \|z_n - x_n\|_1 \leq \frac{1}{m} \|y_n - x_n\|_1 \leq \frac{M}{m} \|y_n - x_n\| \to 0 ,$$

and hence

$$\|z_n - x\| \leq \|z_n - x_n\| + \|x_n - x\| \to 0 .$$

(v)⇒(vi): If $(B + P) \cap (B - P)$ is unbounded, then there exists $\{z_n\} \subset (B + P) \cap (B - P)$ such that $\|z_n\| \to +\infty$, Hence $x_n \leq z_n \leq y_n$, where $x_n, y_n \in B$. Setting $u_n = x_n/\|z_n\|$, $v_n = y_n/\|z_n\|$ and $w_n = z_n/\|z_n\|$, we get $u_n \leq w_n \leq v_n$ and $u_n \to \theta$, $v_n \to \theta$, but $w_n \not\to \theta$ (since $\|w_n\| = 1$), which contradicts (v).

(vi)⇒(vii): Suppose $(B + P) \cap (B - P) \subset \rho B$, where $\rho > 0$. Putting $r = \max\{\|x\|, \|y\|\}$, it is easy to show that $z/r \in (B + P) \cap (B - P)$ for any $z \in [x,y]$; hence $[x,y] \subset r\rho B$.

(vii)$\Rightarrow$(i): Suppose (i) is not true. Then there exist $\{x_n\} \subset P$ and $\{y_n\} \subset P$ such that $\|x_n\| = \|y_n\| = 1$ and $\|x_n + y_n\| < 1/4^n$ $(n = 1,2,3,\ldots)$. Letting

$$u_n = \frac{x_n}{\sqrt{\|x_n + y_n\|}}, \quad v_n = \frac{x_n + y_n}{\sqrt{\|x_n + y_n\|}} \quad (n = 1,2,3,\ldots),$$

we have $\theta \le u_n \le v_n$ and, since

$$\sum_{n=1}^{\infty} \|v_n\| < \sum_{n=1}^{\infty} \frac{1}{2^n} < +\infty ,$$

the series $\sum_{n=1}^{\infty} v_n$ is convergent to some element $v \in E$. Evidently, $\theta \le u_n \le v_n \le v$ and

$$\|u_n\| = \frac{1}{\sqrt{\|x_n + y_n\|}} > 2^n \quad (n = 1,2,3,\ldots);$$

hence, the order interval $[\theta, v]$ is unbounded, which contradicts (vi).

Some authors use the assertion (iii) as the definition of normality of a cone P and call the smallest number N the normal constant of P.

Example 1.1.1. Let $E = R^n$, the Euclidean space, and $P_1 = \{x = (x_1, x_2, \ldots, x_n) \mid x_i \ge 0,\ i = 1,2,\ldots,n\}$. It is clear that P_1 is a cone in R^n and P_1 is solid and generating. Since the norm in R^n is monotone, it follows from Theorem 1.1.1 that P_1 is normal.

Example 1.1.2. Let $E = C(G)$, space of continuous functions on a bounded closed set G in R^n and $P_2 = \{x(t) \in C(G) \mid x(t) \ge 0,\ t \in G\}$. It is easy to see that P_2 is a solid and generating normal cone in $C(G)$. Later we shall use other cones in $C(G)$, such as

$$P_3 = \{x(t) \in C(G) \mid x(t) \ge 0 \text{ and } \int_{G_0} x(t)dt \ge \varepsilon_0 \|x\|_c\},$$

$$P_4 = \{x(t) \in C(G) \mid x(t) \ge 0 \text{ and } \min_{t \in G_0} x(t) \ge \varepsilon_0 \|x\|_c\},$$

where G_0 is a closed subset of G and ε_0 is a given number satisfying

$0 < \varepsilon_0 < 1$. It is easy to show that P_3 and P_4 are solid normal cones in $C(G)$.

Example 1.1.3. Let $E = L^p(\Omega)$, the space of Lebesque measurable functions which are pth power summable on $\Omega \subset R^n$, where $p \geq 1$ and $0 < \text{mes}\ \Omega < +\infty$. Let $P_5 = \{x(t) \in L^p(\Omega) \mid x(t) \geq 0\}$. It is easy to show that P_5 is a generating normal cone in $L^p(\Omega)$, but P_5 is not solid, i.e., $\overset{o}{P_5} = \phi$.

Example 1.1.4. Let $E = C^1[0,2\pi]$, the space of continuously differentiable functions on $[0,2\pi]$ with the norm

$$\|x\| = \max_{0 \leq t \leq 2\pi} |x(t)| + \max_{0 \leq t \leq 2\pi} |x'(t)| ,$$

and let $P_6 = \{x(t) \in C^1[0,2\pi] \mid x(t) \geq 0,\ 0 \leq t \leq 2\pi\}$. Clearly P_6 is a solid cone in $C^1[0,2\pi]$. Also, P_6 is generating, since every $x(t) \in C^1[0,2\pi]$ can be expressed in the form

$$x(t) = y(t) - z(t) ,$$

where $y(t) \equiv M > 0$ and $z(t) = M - x(t)$, $M > \max\limits_{0 \leq t \leq 2\pi} x(t)$, $y(t)$, $z(t) \in P_6$. P_6 is not normal. In fact, if P_6 is normal, then, by Theorem 1.1.1, there exists an $N > 0$ such that $\theta \leq x \leq y \quad \|x\| \leq N\|y\|$. Let $x_n(t) = 1 - \cos nt$, $y_n(t) \equiv 2$. Then we have $\theta \leq x_n \leq y_n$, $\|x_n\| = 2 + n$, and $\|y_n\| = 2$. Consequently, $2 + n \leq 2N$ $(n = 1,2,3,\ldots)$, which is impossible.

1.2 *Regular and Fully Regular Cones.*

Definition 1.2.1. A cone $P \subset E$ is said to be regular if every increasing and bounded in order sequence in E has a limit, i.e., if $\{x_n\}$ E and $y \in E$ satisfy

$$x_1 \leq x_2 \leq \ldots \leq x_n \leq \ldots \leq y , \qquad (1.2.1)$$

then there exists $x^* \in E$ such that $\|x_n - x^*\| \to 0$.

It is clear cone P is regular if and only if every decreasing and bounded in order sequence in E has a limit.

Definition 1.2.2. A cone $P \subset E$ is said to be fully regular if every increasing and bounded in norm sequence in E has a limit, i.e., if $\{x_n\} \subset E$ satisfies

$$x_1 \leq x_2 \leq \ldots \leq x_n \leq \ldots, \quad M = \sup_n \|x_n\| < +\infty, \tag{1.2.2}$$

then there exists $x^* \in E$ such that $\|x_n - x^*\| \to 0$.

It is evident that cone P is fully regular if and only if every decreasing and bounded in norm sequence in E has a limit.

Theorem 1.2.1. Cone P is fully regular $\Rightarrow P$ is regular $\Longrightarrow P$ is normal.

Proof. We first prove that if P is not normal, then P is neither regular nor fully regular. Suppose that P is not normal. By virtue of Theorem 1.1.1 (iii), there exist $\{x_n\} \subset P$ and $\{y_n\} \subset P$ such that $\theta \leq x_n \leq y_n$ and

$$\|x_n\| > 2^n\|y_n\| \quad (n = 1,2,3,\ldots). \tag{1.2.3}$$

Put $z_n = \dfrac{x_n}{\|x_n\|}$ and $v_n = \dfrac{y_n}{2^n\|y_n\|}$ $(n = 1,2,3,\ldots)$, then, by (1.2.3),

$$\theta < z_n \leq \frac{x_n}{2^n\|y_n\|} \leq \frac{y_n}{2^n\|y_n\|} = v_n \quad (n = 1,2,3,\ldots) \tag{1.2.4}$$

and

$$\sum_{n=1}^{\infty} \|v_n\| = \sum_{n=1}^{\infty} \frac{1}{2^n} = 1 < +\infty . \tag{1.2.5}$$

Hence the series $\sum_{n=1}^{\infty} v_n$ converges to some $v \in E$, i.e.,

$$\sum_{n=1}^{\infty} v_n = v. \tag{1.2.6}$$

Now we define

$$w_n = \begin{cases} v_1 + v_2 + \ldots + v_{2m}, & \text{when } n = 2m, \quad m = 1,2,3,\ldots \\ v_1 + v_2 + \ldots + v_{2m} + z_{2m+1}, & \text{when } m = 2m+1, \quad m = 1,2,3,\ldots \end{cases}$$

It is easy to show from (1.2.4), (1.2.5), and (1.2.6) that

$$\theta < w_2 \leq w_3 \leq w_4 \leq w_5 \leq w_6 \leq \ldots \leq v \tag{1.2.7}$$

and

$$\sup_n \|w_n\| \leq 2 < +\infty . \tag{1.2.8}$$

But the sequence w_n does not converge, since $\|w_{2m+1} - w_{2m}\| = \|z_{2m+1}\| = 1$. Hence P is neither regular nor fully regular.

Finally, we need only to prove that the full regularity of P implies the regularity of P. Suppose P is fully regular and (1.2.1) is satisfied. Then, by the conclusion mentioned above, P is normal. It follows therefore from Theorem 1.1.1 and $\theta \leq y - x_n \leq y - x_1$ $(n = 1,2,3,\ldots)$ that

$$\|y - x_n\| \leq N\|y - x_1\| \quad (n = 1,2,3,\ldots) .$$

Hence $\{\|x_n\|\}$ is bounded and x_n converges to some $x^* \in E$.

<u>Example 1.2.1</u>. We prove the cone P_5 in Example 1.1.3 is fully regular (hence, by Theorem 1.2.1, it is regular). Suppose (1.2.2) is satisfied, i.e.,

$$x_1(t) \leq x_2(t) \leq \ldots \leq x_n(t) \leq \ldots, \quad \int_\Omega |x_n(t)|^p \, dt \leq M^p$$
$$(n = 1,2,3,\ldots) .$$

Hence, by Fatou's Lemma, $\int_\Omega |x^*(t)|^p \, dt \leq M$, i.e., $x^* \in L^p(\Omega)$, where $x^*(t) = \lim_{n\to\infty} x_n(t)$. It follows easily from

$$0 \leq x^*(t) - x_n(t) \leq x^*(t) - x\ (t) \quad (n = 1,2,3,\ldots)$$

and the Vitali convergence theorem that

$$\|x^* - x_n\|^p = \int_\Omega |x^*(t) - x_n(t)|^p \, dt \to 0 \quad (n \to \infty).$$

The full regularity of P_5 is proved.

Example 1.2.2. It is easy to prove that the cone P_2 in Example 1.1.2 is not regular (even if it is normal). For simplicity we consider the case $G = [0,1]$. Let $x_n(t) = 1 - t^n$ $(n = 1,2,3,\ldots)$, then

$$x_1(t) \leq x_2(t) \leq \ldots \leq x_n(t) \leq \ldots \leq y(t), \quad y(t) \equiv 1,$$

but x_n does not converge in $C[0,1]$, since $x_n(t)$ does not converge uniformly on $[0,1]$.

Example 1.2.3. Let E be the real Banach space $C_0 = \{x = (x_1,x_2,\ldots,x_k,\ldots) \,|\, x_k \to 0\}$ with the norm $\|x\| = \sup_k |x_k|$ and $P_7 = \{x = (x_1,x_2,\ldots,x_k,\ldots) \in C_0 \,|\, x_k \geq 0,\ k = 1,2,3,\ldots\}$. If $x^{(n)} = (x_1^{(n)},x_2^{(n)},\ldots,x_k^{(n)},\ldots) \in C_0$, $y = (y_1,y_2,\ldots,y_k,\ldots) \in C_0$ such that

$$x^{(1)} \leq x^{(2)} \leq \ldots \leq x^{(n)} \leq \ldots \leq y ,$$

it is easy to see that $\|x^{(n)} - x^*\| \to 0$ $(n \to \infty)$, where

$$x^* = (x_1^*,x_2^*,\ldots,x_k^*,\ldots), \quad x_k^* = \lim_{n\to\infty} x_k^{(n)} .$$

Hence P_7 is regular. On the other hand, putting $z^{(n)} = (z_1^{(n)},z_2^{(n)},\ldots,z_k^{(n)},\ldots)$, where

$$z_k^{(n)} = \begin{cases} 1, & k \leq n; \\ 0, & k > n. \end{cases}$$

we see $z^{(n)} \in C_0$, $z^{(1)} \leq z^{(2)} \leq \ldots \leq z^{(n)} \leq \ldots$ and $\|z^{(n)}\| = 1$ $(n = 1,2,3,\ldots)$, but $\{z^{(n)}\}$ does not converge in C_0 and hence P_7 is not fully regular.

Similarly, we can prove that the cone $P_8 = \{x(t) \in C_0[a,+\infty) \mid x(t) \geq 0\}$ in the real Banach space $C_0[a,+\infty) = \{x = x(t) \mid x(t)$ is continuous on $a \leq t < +\infty$ and $x(t) \to 0$ as $t \to +\infty\}$ with norm

$$\|x\| = \sup_{a \leq t < +\infty} |x(t)|$$

is regular, but is not fully regular.

Theorem 1.2.2. If E is reflexive and P is a cone in E, then the following assertions are equivalent:

(i) P is normal;

(ii) P is regular;

(iii) P is fully regular.

Proof. By virtue of Theorem 1.2.1, we need only prove that the normality of P implies the full regularity of P. Suppose that (1.2.2) is satisfied. We divide the proof into three steps:

(a) It is sufficient to prove that $\{x_n\}$ contains a convergent subsequence $\{x_{n_k}\}$. In fact, if $x_{n_k} \to x^*$ E (i.e., $\|x_{n_k} - x^*\| \to 0$), then, taking $k \to \infty$ in the inequality $x_{n_k} \geq x_m$, where m is any given positive integer and k is sufficiently large, we get $x^* \geq x_m$, and hence

$$x^* \geq x_n \quad (n = 1,2,3,\ldots).$$

By the normality of P, it follows from $\theta \leq x^* - x_n \leq x^* - x_{n_k}$ for $n > n_k$ that $\|x^* - x_n\| \leq N\|x^* - x_{n_k}\|$, and therefore from $\|x^* - x_{n_k}\| \to 0$ $(k \to \infty)$ we know $\|x^* - x_n\| \to 0$ $(n \to \infty)$.

(b) Since E is reflexive and $\{x_n\}$ is bounded, $\{x_n\}$ contains a subsequence $\{x_{n_i}\}$, which converges weakly to $x^* \in E$, i.e., $f(x_{n_i}) \to f(x^*)$ for any $f \in E^*$, where E^* denotes the dual space of E. It must be $x_{n_i} \leq x^*$ $(i = 1,2,3,\ldots)$, since otherwise there exists $x_{n_{i_0}}$ such that $x_{n_{i_0}} \leq x^*$, i.e., $x^* - x_{n_{i_0}} \in P$. By virtue of the second separation theorem of convex sets (see Schaefer [1], p. 65), there exists $f \in E^*$ such that $f(x^* - x_{n_{i_0}}) < c$ and $f(x) > c$ for any $x \in P$, where c is a real number. Consequently,

$$f(x^*) < f(x_{n_{i_0}}) + c \tag{1.2.9}$$

and

$$f(x_{n_i}) > f(x_{n_{i_0}}) + c, \qquad i > i_0. \tag{1.2.10}$$

Taking limit $i \to \infty$ in (1.2.10), we get

$$f(x^*) \geq f(x_{n_{i_0}}) + c$$

which contradicts (1.2.9).

(c) Finally, we prove the sequence $\{x_{n_i}\}$ in Step (b) must contain a subsequence which converges (strongly) to x^*. If it is not the case, then there exist $\varepsilon_0 > 0$ and $m > 0$ such that

$$\|x_{n_i} - x^*\| \geq \varepsilon_0, \qquad i > m.$$

Put $M_i = \{x \in E \mid x \leq x_{n_i}\}$, $M = \bigcup_{i>m} M_i$. Since M_i is convex and $M_i \subset M_{i+1}$ $(i = 1,2,3,\ldots)$, it is easy to see that M is convex and hence $\overline{M}$, the closure of M, is convex. We now prove $x^* \in \overline{M}$. For any $x \in M$, x belongs to some M_i $(i > m)$, i.e., $x \leq x_{n_i}$. By step (b), we have $\theta \leq x^* - x_{n_i} \leq x^* - x$, which implies that $\varepsilon_0 \leq \|x^* - x_{n_i}\| \leq N\|x^* - x\|$. It follows therefore $\|x^* - x\| \geq \varepsilon_0/N$ for every $x \in \overline{M}$, which proves $x^* \notin \overline{M}$. Now, by the second separation theorem of convex sets, there exists $f_0 \in E^*$ and $c_0 \in R^1$ such that $f_0(x^*) < c_0$ and $f_0(x) > c_0$ for any $x \in \overline{M}$. Since $x_{n_i} \in M_i \subset M$, $i > m$, we have $f_0(x_{n_i}) > c_0$ for $i > m$, and therefore

$$f_0(x^*) = \lim_{i\to\infty} f_0(x_{n_i}) \geq c_0,$$

which contradicts $f_0(x^*) < c_0$.

Evidently, the full regularity of P_5 in Example 1.1.3 can be deduced directly from Theorem 1.2.2 when $p > 1$.

Theorem 1.2.3. Let P be a cone in E. P is regular if and only if the following condition holds:

(H_1) $\{x_i\} \subset P$ and $\inf_i \|x_i\| > 0$ imply that $\left\{\sum_{i=1}^{n} x_i\right\}_n$ is unbounded in order, i.e., there does not exist a $z \in E$ such that $\sum_{i=1}^{n} x_i \leq z$, $n = 1,2,3,\ldots$.

Similarly, P is fully regular if and only if the following condition holds:

(H_2) $\{x_i\} \subset P$ and $\inf_i \|x_i\| > 0$ imply $\left\{\sum_{i=1}^{n} x_i\right\}$ is unbounded in norm, i.e., $\sup_n \left\|\sum_{i=1}^{n} x_i\right\| = +\infty$.

Proof. It is easy to see the regularity of P implies (H_1), since, if (H_1) does not hold, then there exist $\{x_i\} \subset P$ and $z \in E$ such that $\inf_i \|x_i\| = a > 0$ and $\sum_{i=1}^{n} x_i \leq z$, $n = 1,2,3,\ldots$. Setting $y_n = \sum_{i=1}^{n} x_i$, we see that

$$y_1 \leq y_2 \leq \cdots \leq y_n \leq \cdots \leq z.$$

But $\|y_{n+1} - y_n\| = \|x_{n+1}\| \geq a$, and hence y_n does not converge, in contradiction with the regularity of P. Similarly, we know the full regularity of P implies (H_2).

Now we prove: if one of the two conditions (H_1) and (H_2) is satisfied, then P must be normal. In fact, if P is not normal, then, by the proof of Theorem 1.2.1, (1.2.3) to (1.2.8) all hold. Let $u_n = w_{n+1} - w_n$ $(n = 2,3,\ldots)$. It is easy to see

$$\{u_n\} \subset P, \qquad \inf_{n=2,3,\ldots} \|u_n\| \geq \frac{13}{16} > 0$$

and

$$\sum_{i=2}^{n} u_i = w_{n+1} - w_2 \leq v - w_2, \quad \|\sum_{i=2}^{n} u_i\| \leq 4, \quad n = 2,3,\ldots,$$

and hence, neither (H_1) nor (H_2) can be satisfied.

Now, suppose (H_1) holds and (1.2.1) is satisfied. We need to prove $\{x_n\}$ is convergent. From the first step (a) in the proof of Theorem 1.2.2, it is sufficient to prove that $\{x_n\}$ contains a convergent subsequence. If $\{x_n\}$ does not contain any convergent subsequence, then set $\{x_1,x_2,x_3,\ldots\}$ is not relatively compact and hence there exist subsequences $\{x_{n_i}\} \subset \{x_n\}$ and $\varepsilon_0 > 0$ such that $\|x_{n_i} - x_{n_j}\| \geq \varepsilon_0$ $(i \neq j)$. Putting $x_i' = x_{n_{i+1}} - x_{n_i}$ $(i = 1,2,3,\ldots)$, we see $\{x_i'\} \subset P$, $\|x_i'\| \geq \varepsilon_0$ $(i = 1,2,3,\ldots)$ and

$$\sum_{i=1}^{m} x_i' = x_{n_{m+1}} - x_{n_1} \leq y - x_{n_1},$$

which contradicts (H_1). Similarly, we can prove (H_2) implies the full regularity of P.

1.3 *Minihedral and Strongly Minihedral Cones.*

An element $z \in E$ is called the least upper bound (i.e., supremum) of set $D \subset E$, if it satisfies two conditions: (i) $x \leq z$ for any $x \in D$; (ii) $x \leq y$, $x \in D$ implies $z \leq y$. We denote the least upper bound of D by $\sup D$, i.e., $z = \sup D$. The greatest lower bound $\inf D$ is defined analogously.

A set $D \subset E$ is said to be bounded above in order if there is an element $w \in E$ such that $x \leq w$ for any $x \in D$. Similarly, we can define a set to be bounded below in order.

Definition 1.3.1. Cone $P \subset E$ is said to be minihedral if $\sup\{x,y\}$ exists for any pair $\{x,y\}$, which is bounded above in order (i.e., $w \in E$ such that $x \leq w, y \leq w$); cone P is said to be strongly minihedral if $\sup D$ exists for any bounded above in order set $D \subset E$.

It is evident, P is minihedral if and only if $\inf\{x,y\}$ exists for any pair $\{x,y\}$, which is bounded below in order. Similarly, P is strongly minihedral if and only if $\inf D$ exists for any bounded below in order set $D \subset E$.

It is easy to show that if P is minihedral, then $\sup D$ exists for any finite set $D = \{x_1,x_2,\ldots,x_n\} \subset E$, which is bounded above in order. Moreover, the following equality holds:

$$\sup\{x_1,x_2,\ldots,x_n\} = \sup\{x_1,\sup\{x_2,\ldots,x_n\}\} .$$

By definition it is clear that the strong minihedrality of a cone P implies the minihedrality of P.

Example 1.3.1. The cone P_2 in Example 1.1.2 is minihedral, since for every $x(t),\ y(t) \in C(G)$, we have $\sup\{x,y\} = z$, where $z(t) = \max\{x(t),y(t)\} \in C(G)$. P_2 is not strongly minihedral, however, since when we let $G = [0,2]$ and $D = \{x(t) \in C[0,2] \mid x(t) < 1$ for $0 < t < 1$ and $x(t) < 2$ for $1 < t < 2\}$, it is easy to see that $\sup D$ does not exist.

Example 1.3.2. The cone P_6 in Example 1.1.4 is not minihedral, since $\sup\{x,y\}$ does not exist for

$$x(t) = t, \quad 0 \leq t \leq 2\pi \quad \text{and} \quad y(t) = 2\pi - t, \quad 0 \leq t \leq 2\pi .$$

It is evident that $x \leq z$, $y \leq z$, where $z(t) = 2\pi$, $0 \leq t \leq 2\pi$.

On the other hand, the cone P_7 in Example 1.2.3 is strongly minihedral, since $\sup D = z \in c_0$ for every bounded above in order set $D \subset c_0$, where

$$z = (z_1,z_2,\ldots,z_k,\ldots)$$

and

$$z_k = \sup\{x_k \mid x = (x_1,x_2,\ldots,x_k,\ldots) \in D\}, \quad k = 1,2,3,\ldots .$$

Theorem 1.3.1. If E is separable and the cone $P \subset E$ is regular and minihedral, then P is strongly minihedral.

Proof. Suppose $D \subset E$ and there is a $z \in E$ such that $x \leq z$, $\forall\, x \in D$. Since E is separable, there exists a denumerable set $M = \{x_1,x_2,\ldots,x_n,\ldots\} \subset D$, which is dense in D. Denote $y_n = \sup\{x_1,x_2,\ldots,x_n\}$ $(n = 1,2,3,\ldots)$, which exist by virtue of the minihedrality of P. Evidently,

$$y_1 \leq y_2 \leq \cdots \leq y_n \leq \cdots \leq z .$$

It follows from the regularity of P that $y_n \to x^* \in E$ (i.e., $\|y_n - x^*\| \to 0$).

Now, we prove $x^* = \sup D$. First, we notice

$$x_n \leq x^* \quad (n = 1,2,3,\ldots). \tag{1.3.1}$$

For any $x \in D$, there exists a sequence in M, which converges to x. Observing (1.3.1), we get $x \leq x^*$, hence x^* is an upper bound of D. On the other hand, if $v \in E$ is any upper bound of D, then

$$y_n \leq v \quad (n = 1,2,3,\ldots)$$

and therefore $x^* \leq v$. □

Corollary 1.3.1. If E is separable and reflexive and the cone $P \subset E$ is normal and minihedral, then P is strongly minihedral.

Proof. The required conclusion follows directly from Theorems 1.2.2 and 1.3.1.

Example 1.3.1. The cone P_5 in Example 1.1.3 is minihedral, since $\sup\{x,y\} = z$ exists for every $x(t)$, $y(t) \in L^p(\Omega)\,(p \geq 1)$, where

$z(t) = \max\{x(t), y(t)\} \in L^p(\Omega)$. Observing that $L^p(\Omega)$ is separable and P_5 is regular, as proven in Example 1.2.1, it follows from Theorem 1.3.1 that P_5 is strongly minihedral.

Example 1.3.4. The cone P_6 in Example 1.1.4 is neither minihedral nor normal. Now, we give an example of a cone which is minihedral, but is not normal. Let $E = \ell^2$ with norm

$$\|x\| = \left(\sum_{i=1}^{\infty} x_i^2 \right)^{\frac{1}{2}}, \quad x = (x_1, x_2, \ldots, x_i, \ldots).$$

Let

$$P_8 = \{x = (x_1, x_2, \ldots, x\ , \ldots) \in \ell^2 \mid x_1 \geq 0,\quad x_i \leq x_1,\quad i = 2,3,\ldots\}.$$

Obviously, P_8 is a cone in ℓ^2. For every pair of elements $x = (x_1, x_2, \ldots, x_i, \ldots) \in \ell^2$ and $y = (y_1, y_2, \ldots, y_i, \ldots) \in \ell^2$, it is easy to verify that $\sup\{x,y\} = z$, where $z = (z_1, z_2, \ldots, z_i, \ldots)$,

$$z_1 = \max\{x_1, y_1\},$$

$$z_i = \min\{z_1 - x_1 + x_i,\ z_1 - y_1 + y_i\}, \quad i = 2,3,\ldots\ .$$

In case $z_1 = x_1$, we have $z_i = x_i$ or $y_i \leq z_i \leq x_i$, and in case $z_1 = y_1$, we have $z_i = y_i$ or $x_i \leq z_i \leq y_i$; hence we must have $z \in \ell^2$, and the minihedrality of P_8 has been proven.

Now, let $u = (1, 1/2, 1/3, \ldots, 1/i, \ldots)$, $\theta = (0,0,\ldots,0,\ldots)$, and $x^{(n)} = (x_1^{(n)}, x_2^{(n)}, \ldots, x_i^{(n)}, \ldots)$, where

$$x_i^{(n)} = \begin{cases} \frac{1}{2}, & i \leq n; \\ 0, & i > n. \end{cases}$$

It is easy to see that $\theta \leq x^{(n)} \leq u$ $(n = 1,2,3,\ldots)$, $\|x^{(n)}\| = \sqrt{n}/2 \to +\infty$, and hence $[\theta, u]$ is unbounded. Consequently by Theorem 1.1.1, P_8 is not normal.

We can also give a simple example of a cone which is strongly minihedral and normal, but is not regular. Let E be the space m of all bounded sequences $x = (x_1, x_2, \ldots, x_i, \ldots)$ with norm $\|x\| = \sup_i |x_i|$ and let $P_9 = \{x = (x_1, x_2, \ldots, x_i, \ldots) \in m \mid x_i \geq 0, \quad i = 1, 2, 3, \ldots\}$. Since the norm is monotone, P_9 is a normal cone in m. Let D be a bounded above in order set in m. It is clear $\sup D = z \in m$, where $z = (z_1, z_2, \ldots, z_i, \ldots)$,

$$z_i = \sup\{x_i \mid x = (x_1, x_2, \ldots, x_i, \ldots) \in D\}.$$

Hence P_9 is strongly minihedral. Finally, the sequence $\{x^{(n)}\}$, $x^{(n)} = (x_1^{(n)}, x_2^{(n)}, \ldots, x_i^{(n)}, \ldots)$, where

$$x_i^{(n)} = \begin{cases} 1, & i \leq n; \\ 0, & i > n, \end{cases}$$

satisfies

$$x^{(1)} \leq x^{(2)} \leq \ldots \leq x^{(n)} \leq \ldots \leq y,$$

with $y = (1, 2, \ldots, 1, \ldots)$. But $x^{(n)}$ does not converge in m and hence P_9 is not regular.

1.4 *Positive Linear Functionals.*

Let E be a real Banach space and P be a cone in E. Functional $f \in E^*$ is said to be <u>positive</u> if $f(x) \geq 0$ for any $x \in P$. The set of all positive bounded linear functionals is denoted by P^*, i.e.,

$$P^* = \{f \in E^* \mid f(x) \geq 0, \quad x \in P\}. \tag{1.4.1}$$

It is easy to see that P^* satisfies all conditions of a cone except

$$P^* \cap (-P^*) = \{\theta\}, \tag{1.4.2}$$

and hence, if P is generating, then (1.4.2) is satisfied and P^* is a cone in E^*, which is called the dual cone of P.

Theorem 1.4.1. The following conclusions hold:

(i) $x \in P$ if and only if $f(x) \geq 0$ for any $f \in P^*$; for $x_1 > \theta$ there exists $f_1 \in P^*$ such that $f_1(x_1) > 0$; for $x_2 \in P$, there exists $f_2 \in P^*$ such that $f_2(x_2) < 0$.

(ii) Let P be solid. Then $x \in \overset{o}{P}$ if and only if $f(x) > 0$ for any $f \in P^* \backslash \{\theta\}$.

(iii) If E is separable, then there exists $f_0 \in P^*$ such that $f_0(x) > 0$ for any $x > \theta$.

Proof. We first prove (i). It is evident that we need only to prove the last conclusion in (i). For a given $x_2 \in P$, by the second separation theorem of convex sets, there exists $f_2 \in E^*$ such that $f_2(x_2) < c$ and $f_2(x) > c$ for any $x \in P$, where c is a real number. Since $\theta \in P$, we have $c < f_2(\theta) = 0$ and therefore $f_2(x_2) < 0$. For any $x \in P$, we have $nx \in P$, $n = 1,2,3,\ldots$, and so $f_2(nx) > c$, i.e.,

$$f_2(x) > \frac{c}{n}, \quad n = 1,2,3,\ldots .$$

Letting $n \to \infty$, we get $f_2(x) \geq 0$ and hence $f_2 \in P^*$.

Next we prove (ii). Suppose $x_0 \in \overset{o}{P}$. There exists $r > 0$ such that $S(x_0,r) = \{x \in E \mid \|x - x_0\| \leq r\} \subset P$ and so $x_0 \pm rz \geq \theta$, $\forall\ z \in E$, $\|z\| \leq 1$. Hence, for any $f \in P^* \backslash \{\theta\}$, $f(x_0 \pm rz) \geq 0$, and therefore $f(x_0) \geq r|f(z)|$, $\forall\ \|z\| \leq 1$. It follows that $f(x_0) \geq r\|f\| > 0$.

Now we suppose $x_0 \notin \overset{o}{P}$. Putting $P_1 = \{x = tx_0 \mid t \geq 0\}$, we see P_1 is a cone in E and $\overset{o}{P} \cap P_1 = \phi$. By virtue of the first separation theorem of convex sets (see Schaefer [1], p. 64), there exist $f_0 \in E^*$ and $c_0 \in R'$ such that $f_0 \neq \theta$, and

$$f_0(x) \leq c_0, \quad x \in P_1; \quad f_0(x) \geq c_0, \quad x \in P.$$

Since $\theta \in P_1 \cap P$, we know $c_0 = 0$, and so $f_0 \in P^*\backslash\{\theta\}$, $f_0(x_0) \leq 0$.

Finally, we prove (iii). Since E is separable, there exists a denumerable set $F = \{x_1, x_2, \ldots, x_n, \ldots\} \subset B = \{x \in E \mid \|x\| \leq 1\}$, which is dense in B. Define a new metric in E^* by

$$d(f_1, f_2) = \sup_n \frac{f_1(x_n) - f_2(x_n)}{n}, \quad f_1, f_2 \in E^* . \tag{1.4.3}$$

It is easy to verify that under metric (1.4.3) E^* becomes a metric space, and therefore $B^* = \{f \in E^* \mid \|f\| \leq 1\}$ is also a metric space. Evidently, if $d(f_m, f_0) \to 0$ $(f_m, f_0 \in B^*)$, then g_m w^*-converges to f_0, i.e., $f_m(x) \to f_0(x)$ $(m \to \infty)$ for any $x \in E$; conversely, if f_m w^*-converges to f_0 $(f_m, f_0 \in B^*)$, then

$$M = \sup\{\|f_0\|, \|f_1\|, \ldots, \|f_m\|, \ldots\} < +\infty .$$

For any $\varepsilon > 0$, choose a positive integer n_0 such that

$$\frac{2M}{n} < \frac{\varepsilon}{2}, \quad \forall\, n > n_0.$$

As a result, we have

$$\frac{|f_m(x_n) - f_0(x_n)|}{n} \leq \frac{2M}{n} < \frac{\varepsilon}{2}, \quad \forall\, n > n_0, \quad m = 1,2,3,\ldots . \tag{1.4.4}$$

Now, choose a positive integer m_0 such that

$$\frac{|f_m(x_n) - f_0(x_n)|}{n} < \frac{\varepsilon}{2}, \quad \forall\, n = 1,2,\ldots,n_0, \quad m > m_0. \tag{1.4.5}$$

It follows from (1.4.4) and (1.4.5) that

$$d(f_m, f_0) = \sup_n \frac{|f_m(x_n) - f_0(x_n)|}{n} \leq \varepsilon, \quad \forall\, m > m_0,$$

and hence $d(f_m, f_0) \to 0$.

Since E is separable, B^* is w^*-compact and w^*-closed, and therefore $R^* = B^* \cap P^*$ is a compact metric space under metric (1.4.3).

Hence R^* is separable, i.e., there exists a denumerable set $V = \{f_1, f_2, \ldots, f_n, \ldots\} \subset R^*$, which is dense (under metric (1.4.3)) in R^*. Now, put $f_0 = \sum_{n=1}^{\infty} f_n/2^n$, i.e.,

$$f_0(x) = \sum_{n=1}^{\infty} \frac{f_n(x)}{2^n}, \quad \forall\, x \in E.$$

It is easy to see that $f_0 \in P^*$. We prove $f_0(x) > 0$, $\forall\, x > \theta$. In fact, if $f_0(x_0) = 0$ for some $x_0 > \theta$, then $f_n(x_0) = 0$ $(n = 1,2,3,\ldots)$. On account of the density of V in R^* and the equivalence of metric (1.4.3) convergence and w^*-convergence, we see that $f(x_0) = 0$, $\forall\, f \in R^*$ and therefore, $f(x_0) = 0$ for any $f \in P^*$, which contradicts conclusion (i).

Lemma 1.4.1. If P is solid, then P is generating.

Proof. Let $x_0 \in \overset{o}{P}$. There is a $r > 0$ such that $S(x_0, r) = \{x \in E \mid \|x - x_0\| \leq r\} \subset P$.

For any $x \in E$, we have $x_0 \pm r/\|x\|\, x \in S(x_0, r) \subset P$. Hence $x = y - z$, where

$$y = \frac{\|x\|}{2r}\left(x_0 + \frac{r}{\|x\|}\, x\right) \in P, \quad z = \frac{\|z\|}{2r}\left(x_0 - \frac{r}{\|x\|}\, x\right) \in P.$$

Lemma 1.4.2. If P is generating, then there exists a $\gamma > 0$ such that

$$\gamma B \subset B \cap P - B \cap P, \tag{1.4.6}$$

where $B = \{x \in E \mid \|x\| \leq 1\}$, i.e., every element $x \in E$ can be represented in the form $x = u - v$, where $u, v \in P$ and

$$\|u\| \leq \tau \|x\|, \quad \|v\| \leq \tau \|x\|, \quad \tau = \gamma^{-1}. \tag{1.4.7}$$

Proof. We need only prove that there exists a $\sigma > 0$ such that every element $x \in E$ corresponds to $u \in P$, which satisfies $x \leq u$ and $\|u\| \leq \sigma \|x\|$. In fact, in this case, (1.4.7) is satisfied with $v = u - x$ and $\tau = \sigma + 1$.

Since P is generating, we have $E = \sum_{n=1}^{\infty} E_n$, where

$$E_n = \{x \in E \mid \exists\, u \in P,\ x \le u \text{ and } \|u\| \le n\|x\|\} \quad (n = 1,2,3,\ldots).$$

By the Baire category theorem, there exist positive integer n_1 and $x_0 \in E$, $R > r > 0$ such that $\overline{E}_{n_1} \supset T = \{x \in E \mid r \le \|x - x_0\| \le R\}$. Let $-x_0 = y_0 - z_0$, where $y_0, z_0 \in P$. Then $-x_0 \le y_0$. Choose a positive integer n_2 such that $\|y_0\| \le n_2\|x_0\|$. For any $x \in E_{n_1} \cap T$, there exists $u \in P$ such that $x \le u$ and $\|u\| \le n_1\|x\|$. Hence $x - x_0 \le u + y_0 \in P$ and

$$\|u + y_0\| \le n_1\|x\| + n_2\|x_0\| \le (n_1 + n_2)\|x_0\| + n_1\|x - x_0\| \le n_3\|x - x_0\| ,$$

where $n_3 = [(n_1 + n_2)\|x_0\|/r + n_1] + 1$, and so $x - x_0 \in E_{n_3}$. It follows therefore $\overline{E}_{n_3} \supset T_0 = \{x \in E \mid r \le \|x\| \le R\}$. Since $x \in E_{n_3}$ implies $tx \in E_{n_3}$ for $t \ge 0$, we get $\overline{E}_{n_3} = E$.

Now we prove $E = E_{3n_3}$. For a given $x \in E$ $(x \ne \theta)$, there exists $x_1 \in E_{n_3}$ such that

$$\|x - x_1\| < \frac{1}{2}\|x\|.$$

Since $x_1 \in E_{n_3}$, there is a $u_1 \in P$ such that

$$x_1 \le u_1, \quad \|u_1\| \le n_3\|x_1\| .$$

Similarly, there exist $x_2 \in E_{n_3}$ and $u_2 \in P$ such that

$$\|x - x_1 - x_2\| < \frac{1}{2^2}\|x\|, \quad x_2 \le u_2 , \quad \|u_2\| \le n_3\|x_2\| .$$

Continuing this process, we get sequences $\{x_k\} \subset E$ and $\{u_k\} \subset P$ such that

$$\|x - \sum_{i=1}^{k} x_i\| < \frac{1}{2^k}\|x\|, \quad x_k \le u_k ,$$

$$\|u_k\| \le n_3\|x_k\|, \quad k = 1,2,3,\ldots .$$

Evidently, $x = \sum_{k=1}^{\infty} x_k$ and

$$\|x_k\| \le \|x - \sum_{i=1}^{k-1} x_i\| + \|x - \sum_{i=1}^{k} x_i\| < \frac{3\|x\|}{2^k},$$

and so

$$\sum_{k=1}^{\infty} \|u_k\| \le n_3 \sum_{k=1}^{\infty} \|x_k\| \le 3n_3\|x\| < +\infty .$$

Hence, the series $\sum_{k=1}^{\infty} u_k$ converges to some $u \in P$ and

$$x = \sum_{k=1}^{\infty} x_k \le \sum_{k=1}^{\infty} u_k = u,$$

$$\|u\| \le \sum_{k=1}^{\infty} \|u_k\| \le 3n_3\|x\| .$$

It follows therefore $x \in E_{3n_3}$ and $E = E_{3n_3}$ is proved. Thus, the conclusion in the beginning of the proof holds by setting $\sigma = 3n_3$. □

Theorem 1.4.2. If P is generating, then P^* is a normal cone in E^*.

Proof. Let $f,g \in E^*$ and $\theta \le f \le g$. By Lemma 1.4.2, any $x \in E$ can be represented in the form $x = u - v$, where $u,v \in P$ and (1.4.7) holds with $\gamma > 0$, a constant. We have

$$f(x) = f(u) - f(v) \le f(u) \le g(u) \le \|g\|\cdot\|u\| \le \gamma^{-1}\|g\|\cdot\|x\| ,$$

and

$$-f(x) = f(-x) = f(v) - f(u) \le f(v) \le g(v) \le \|g\|\cdot\|v\| \le \gamma^{-1}\|g\|\cdot\|x\| .$$

Hence

$$|f(x)| \le \gamma^{-1}\|g\|\cdot\|x\| ,$$

and so

$$\|f\| \le \gamma^{-1}\|g\| .$$

Thus, the normality of P^* follows from Theorem 1.1.1. □

Corollary 1.4.1. If P is solid, then P^* is normal.

Finally, we give an extension theorem of positive bounded linear functionals.

Theorem 1.4.3. Let P be solid and G be a linear subspace of E such that $\overset{o}{P} \cap G \neq \phi$. Suppose that f is a positive bounded linear functional on G, i.e., $f(x) \geq 0$ for any $x \in G \cap P$. Then f can be extended to a positive bounded linear functional on E, i.e., there exists $\tilde{f} \in P^*$ such that $\tilde{f}(x) = f(x)$ for any $x \in G$.

Proof. Let $F = \{G_1 \supset G \mid G_1$ is a linear subspace of E and f can be extended to a positive bounded linear functional f_1 on $G_1\}$. For every $G_1 \in F$, we fix a f_1 corresponding to G_1. Evidently, F is partially ordered by defining $G_1 < G_2$ $(G_1, G_2 \in F)$ to mean that $G_1 \subset G_2$ and f_2 is an extension of f_1, where f_1 and f_2 correspond to G_1 and G_2, respectively. Now, suppose that H is a completely ordered subset of F. Putting

$$G_H = \bigcup_{G_1 \in H} G_1 ,$$

we see that $G_H \supset G$ and G_H is a linear subspace of E. Define a functional f_H on G_H as follows: For any $x \in G_H$, there exists $G_1 \in H$ such that $x \in G_1$, and let $f_H(x) = f_1(x)$, where f_1 corresponds to G_1. It is easy to show that $f_H(x)$ is independent of the choice of G_1 and f_H is a positive linear functional on G_H. Now, we prove the boundness of f_H. Let $x_0 \in \overset{o}{P} \cap G$, then there is $r > 0$ such that $S(x_0, r) = \{x \in E \mid \|x - x_0\| \leq r\} \subset P$. For any $z \in G_H$ satisfying $\|z\| \leq 1$, we have $x_0 \pm rz \in G_H \cap P$, and so $f_H(x_0 \pm rz) \geq 0$ or $f_H(x_0) \geq \pm r f_H(z)$. Hence

$$|f_H(z)| \leq \frac{1}{r} f_H(x_0) = \frac{1}{r} f\ (x_0) = \text{const.}, \tag{1.4.8}$$

which shows that f_H is bounded and $\|f_H\| \leq 1/r\ f(x_0)$. Thus, $f_H \in F$ and f_H is an upper bound of H. It follows therefore from Zorn's Lemma that F contains a maximal element G_m. We now prove $G_m = E$. If this is not true, then there exists $y_0 \in E \backslash G_m$. As mentioned above,

$S(x_0, r) \subset P$, $x_0 \in \overset{o}{P} \cap G \subset \overset{o}{P} \cap G_m$. Hence $x_0 \pm r/\|y_0\| \, y_0 \geq \theta$ and

$$x_1 \leq y_0 \leq x_2 \,, \tag{1.4.9}$$

where $x_1 = -\|y_0\|/r \, x_0 \in G_m$ and $x_2 = \|y_0\|/r \, x_0 \in G_m$. Denoting the positive bounded linear extension of f on G_m by f_m and observing (1.4.9), we can choose a number ξ such that

$$\sup_{x' \leq y_0, x' \in G_m} f_m(x') \leq \xi \leq \inf_{x'' \geq y_0, x'' \in G_m} f_m(x'') \,. \tag{1.4.10}$$

Put

$$G_* = \{y = x + ty_0 \mid x \in G_m, \quad t \in R^1\}.$$

Evidently, G_* is a linear subspace of E and $G_* \supset G_m$, $G_* \neq G_m$. Now we define

$$f_*(y) = f_m(x) + t\xi, \quad \forall \; y = x + ty_0 \in G_*. \tag{1.4.11}$$

It is easy to see that the expression $y = x + ty_0$ is unique for every $y \in G_*$; hence $f_*(y)$ is defined uniquely. Clearly, the functional f_* is an extension of f_m on G_* and f_* is additive; hence

$$f_*(\lambda y) = \lambda f_*(y), \quad \forall \; y \in G_*, \quad \lambda \text{ is any rational number.} \tag{1.4.12}$$

We now prove

$$f_*(y) \geq 0, \quad y \in G_* \cap P. \tag{1.4.13}$$

In fact, for any $y = x + ty_0 \geq \theta$ $(x \in G_m, t \in R^1)$, we have

$$y_0 \geq -\frac{x}{t} \,, \quad t > 0; \quad y_0 \leq -\frac{x}{t} \,, \quad t < 0 \,.$$

It follows therefore from (1.4.10) that

$$\xi \geq -\frac{1}{t} f_m(x), \quad \forall\, t > 0; \quad \xi \leq -\frac{1}{t} f_m(x), \quad \forall\, t < 0,$$

and therefore,

$$f_*(y) = f_m(x) + t\xi \geq 0, \quad \forall\, t \in R^1,$$

which proves (1.4.13). Next we prove

$$f_*(z_n) \to 0 \quad \text{if} \quad z_n \in G_* \quad (n = 1,2,3,\dots) \quad \text{and} \quad \|z_n\| \to 0. \tag{1.4.14}$$

In fact, in a way similar to the proof of (1.4.8), we can get

$$|f_*(z)| \leq \frac{1}{r} f(x_0), \quad \forall\, z \in G_*, \quad \|z\| \leq 1. \tag{1.4.15}$$

For arbitrary $\varepsilon > 0$, we first choose a positive integer k such that

$$\frac{1}{kr} f(x_0) < \varepsilon,$$

and then choose positive integer N such that $\|z_n\| < 1/k$ for $n > N$. It follows from (1.4.12) and (1.4.15) that

$$|kf_*(z_n)| = |f_*(kz_n)| \leq \frac{1}{r} f(x_0), \quad \forall\, n > N,$$

and hence

$$|f_*(z_n)| \leq \frac{1}{kr} f(x_0) < \varepsilon, \ \forall\, n > N,$$

which proves (1.4.14). Finally, we prove (1.4.12) is still true for irrational number λ. Let irrational number λ be given. Choosing a sequence of rational numbers $\{\lambda_n\}$, which tends to λ, we get from (1.4.12) and (1.4.14) that for any $y \in G_*$,

$$\begin{aligned} |f_*(\lambda y) - \lambda f_*(y)| &\leq |f_*(\lambda y) - f_*(\lambda_n y)| + |\lambda_n - \lambda|\cdot|f_*(y)| \\ &= |f_*(\lambda - \lambda_n)y)| + |\lambda_n - \lambda|\cdot|f_*(y)| \to 0 \quad (n \to \infty). \end{aligned}$$

Thus (1.4.12) holds for a given irrational number λ. It follows therefore f_* is linear and by (1.4.15) f_* is bounded. Consequently, $G_* \in F$, which contradicts the maximality of F. □

1.5 The e-Norm and Hilbert's Projective Metric.

(a) The e-Norm

Let P be a cone in real Banach space E and $e > \theta$. Set

$$E_e = \{x \in E \mid \text{there exists } \lambda > 0 \text{ such that } -\lambda e \leq x \leq \lambda e\}$$

and

$$\|x\|_e = \inf\{\lambda > 0 \mid -\lambda e \leq x \leq \lambda e\}, \ \forall \ x \in E_e. \tag{1.5.1}$$

It is easy to see that E_e becomes a normed linear space under the norm $\|\cdot\|_e$. $\|x\|_e$ is called the e-norm of the element $x \in E_e$.

Theorem 1.5.1. Let cone P be normal. Then the following conclusions hold:

(i) E_e is a Banach space with e-norm, and there exists a constant $M > 0$ such that $\|x\| \leq M\|x\|_e$ for any $x \in E_e$;

(ii) $P_e = E_e \cap P$ is a normal solid cone of E_e;

(iii) if P is solid and $e \in \overset{o}{P}$, then $E_e = E$ and the e-norm $\|\cdot\|_e$ is equivalent to the original norm $\|\cdot\|$.

Proof. We first prove (i). For $x \in E_e$, we have $-\alpha e \leq x \leq \alpha e$, where $\alpha = \|x\|_e$, and so $\theta \leq x + \alpha e \leq 2\alpha e$. Hence, by the normality of P and Theorem 1.1.1, $\|x + \alpha e\| \leq N\| 2\alpha e\|$, where N is the normal constant of P. It follows therefore

$$\|x\| \leq \|x + \alpha e\| + \|-\alpha e\| \leq M\|x\|_e , \tag{1.5.2}$$

where $M = (2N + 1)\|e\| = \text{const}$.

Now we prove the completeness of E_e. Let $\{x_n\}$ be a fundamental sequence of E_e, i.e., $\|x_n - x_m\|_e \to 0$ $(n, m \to \infty)$. By (1.5.2), we have

$$\|x_n - x_m\| \leq M\|x_n - x_m\|_e ,$$

and hence $\|x_n - x_m\| \to 0$ $(n, m \to \infty)$. By the completeness of E, there exists an $x^* \in E$ such that $\|x_n - x^*\| \to 0$ $(n \to \infty)$. For any given $\varepsilon > 0$, there exists a positive integer N_1 such that $\|x_n - x_m\|_e \leq \varepsilon$, i.e.,

$$-\varepsilon e \leq x_n - x_m \leq \varepsilon_e, \quad \forall\ n > N_1, \quad m > N_1. \tag{1.5.3}$$

Let n be fixed and $m \to \infty$ in (1.5.3), then we get

$$-\varepsilon e \leq x_n - x^* \leq \varepsilon e, \quad \forall\ n > N_1,$$

which implies $\|x_n - x^*\|_e \leq \varepsilon$, $\forall\ n > N_1$. It follows therefore $x^* \in E_e$ and $\|x_n - x^*\| \to 0$ $(n \to \infty)$ and the completeness of E_e is proved.

Next we prove (ii). It is easy to see that $x \in E_e$ belongs to $\overset{o}{P}_e$ if and only if $x \geq \tau e$ for some $\tau > 0$. Hence $e \in \overset{o}{P}_e$, which shows that P_e is solid. The normality of P_e is obvious, since the e-norm is monotone, i.e., $\theta \leq x \leq y, x, y \in E_e$ implies $\|x\|_e \leq \|y\|_e$.

Finally, we prove conclusion (iii). Since $e \in \overset{o}{P}$, there is a $r > 0$ such that $S(e,r) = \{x \in E \mid \|x - e\| \leq r\} \subset P$. For any $x \in E$, we have $e \pm r/\|x\|\ x \in P$, and therefore

$$-\frac{x}{r}\, e \leq x \leq \frac{x}{r}\, e .$$

Thus $x \in E_e$ and

$$\|x\|_e \leq \frac{1}{r}\, \|x\|, \quad \forall\, x \in E. \tag{1.5.4}$$

It follows $E_e = E$, and observing (1.5.4) and conclusion (i), $\|\cdot\|_e$ is equivalent to $\|\cdot\|$. □

Theorem 1.5.1 shows that from a normal cone P we can construct another normal cone P_e, which is solid. Sometimes this is convenient in applications as in Chapter 2.

Example 1.5.1. Consider P_5 in Example 1.1.3. Let $e(t) \equiv 1$, $\forall\, t \in \Omega$. Then $e \in P_5$, $e > \theta$. It is easy to show that $E_e = L^\infty(\Omega)$-space of bounded measurable functions on Ω and

$$\|x\|_e = \|x\|_{L^\infty} = \operatorname*{vrai\,max}_{t\in\Omega} |x(t)| \,. \tag{1.5.5}$$

Since P_5 is normal, it follows from Theorem 1.5.1 that there exists a constant $M > 0$ such that

$$\|x\| \le M\|x\|_e, \quad \forall\, x \in E_e = L^\infty(\Omega).$$

Notice that this inequality can be deduced from (1.5.5) directly:

$$\|x\|_{L^p} \le (\text{mes } \Omega)^{\frac{1}{p}} \cdot \|x\|_{L^\infty} \,.$$

Example 1.5.2. Let $E = C[0,1]$ and $P_{10} = \{x(t) \in C[0,1] \mid x(t) \ge 0,\ 0 \le t \le 1\}$. First we set $e(t) = 1/2\, t(1 - t)$, i.e., $e(t)$ is the solution of the two-point boundary value problem

$$x'' = -1, \quad x(0) = x(1) = 0.$$

It is evident that $e > \theta$ and

$$E_e = \{x(t) \in C[0,1] \mid \exists\, a > 0,\ |x(t)| \le \tfrac{a}{2}\, t(1 - t),\quad t \in [0,1]\}, \tag{1.5.6}$$

$$\|x\|_e = \sup_{0<t<1} \frac{2|x(t)|}{t(1 - t)}, \quad \forall\, x \in E_e, \tag{1.5.7}$$

$$(P_{10})_e = \{x(t) \in C[0,1] \mid \exists\, a > 0,\ 0 \le x(t) \le \tfrac{a}{2}\, t(1 - t),\ \forall\, t \in [0,1]\}, \tag{1.5.8}$$

$$(\overset{o}{P}_{10})_e = \{x(t) \in C[0,1] \mid \exists\ a > b > 0,$$
$$\frac{b}{2}\, t(1-t) \le x(t) \le \frac{a}{2}\, t(1-t),\ \forall\ t \in [0,1]\}\ . \tag{1.5.9}$$

Since P_{10} is normal, by Theorem 1.5.1 we know that E_e is a Banach space with norm (1.5.7) and there exists a constant $M > 0$ such that

$$\|x\| \le M\|x\|_e\ ,\ \forall\ x \in E_e\ ; \tag{1.5.10}$$

moreover, $(P_{10})_e$ is a normal solid cone of E_e. We must point out that (1.5.10) can be deduced from (1.5.7) directly:

$$\|x\|_C = \max_{0\le t\le 1} |x(t)| \le \|x\|_e \cdot \left(\max_{0\le t\le 1} \frac{1}{2}\, t(1-t)\right) = \frac{1}{8}\,\|x\|_e\ .$$

Next, we set $e(t) \equiv 2$, $t \in [0,1]$. Evidently, $e \in \overset{o}{P}_{10}$ and $E_e = E = C[0,1]$. It is easy to see that

$$\|x\|_C = 2\|x\|_e$$

for any $x \in E_e = E$ and hence the norm $\|\cdot\|_e$ is equivalent to the norm $\|\cdot\|_e$, by Theorem 1.5.1.

(b) Hilbert's Projective Metric

Let P be a solid cone in real Banach space E. For given $x,y \in \overset{o}{P}$, there exist sufficiently small positive number μ and sufficiently large positive number λ such that $x - \mu y \in P$ and $y - 1/\lambda\ x \in P$, i.e., $\mu y \le x \le \lambda y$. Hence, we can define

$$m\left(\frac{x}{y}\right) = \sup\{\mu > 0 \mid \mu y \le x\},\quad M\left(\frac{x}{y}\right) = \inf|\lambda > 0 \mid x \le \lambda y\}.$$

As a result, we have

$$0 < m\left(\frac{x}{y}\right) \le M\left(\frac{x}{y}\right) \quad\text{and}\quad m\left(\frac{x}{y}\right)y \le x \le M\left(\frac{x}{y}\right)y\ .$$

The Hilbert's projective metric is then defined by

$$d(x,y) = ln\left\{\frac{M\left(\frac{x}{y}\right)}{m\left(\frac{x}{y}\right)}\right. . \tag{1.5.11}$$

Lemma 1.5.1. $d(x,y)$ is a quasi-metric in $\overset{o}{P}$, i.e., $d(x,y)$ satisfies the following three conditions:

(i) $d(x,y) = 0, \ \forall\, x \in \overset{o}{P}$;

(ii) $d(x,y) = d(y,x), \ \forall\, x,y \in \overset{o}{P}$;

(iii) $d(x,y) \leq d(x,z) + d(z,y), \ \forall\, x,y,z \in \overset{o}{P}$.

Moreover, we have

(iv) $d(\lambda x,\mu y) = d(x,y), \ \forall\, x,y \in \overset{o}{P}, \ \lambda > 0, \ \mu > 0$;

(v) $d(x,y) = 0$ if and only if $x = \lambda y$, where $\lambda > 0$.

Proof. The proof is trivial and we only point out that in the proof of (iii) we must use the inequalities

$$M\left(\frac{x}{y}\right) \leq M\left(\frac{x}{z}\right)M\left(\frac{z}{y}\right)$$

and

$$m\left(\frac{x}{y}\right) \geq m\left(\frac{x}{z}\right)m\left(\frac{z}{y}\right) ,$$

which can be deduced from the following inequalities:

$$m\left(\frac{x}{z}\right)m\left(\frac{z}{y}\right)y \leq m\left(\frac{x}{z}\right)z \leq x \leq M\left(\frac{x}{z}\right)z \leq M\left(\frac{x}{z}\right)M\left(\frac{z}{y}\right)y \ . \ \square$$

From Lemma 1.5.1 we know that $(\overset{o}{P} \cap S_1,d)$ is a metric space, where $S_1 = \{x \in E \mid \|x\| = 1\}$.

Lemma 1.5.2. If the norm of E is monotone with respect to P, i.e., $\theta \leq x \leq y$ implies $\|x\| \leq \|y\|$, then the following two conclusions hold:

(i) $0 \leq m\left(\frac{x}{y}\right) \leq 1 \leq M\left(\frac{x}{y}\right), \ \forall\, x,y \in \overset{o}{P} \cap S_1$;

(ii) $\|x - y\| \leq 2\{e^{d(x,y)} - 1\}, \ \forall\, x,y \in \overset{o}{P} \cap S_1$,

where $S_1 = \{x \in E \mid \|x\| = 1\}$.

Proof. We first prove (i). For $x,y \in \overset{o}{P} \cap S_1$, we have

$$m\left(\frac{x}{y}\right)y \leq x \leq M\left(\frac{x}{y}\right)y , \tag{1.5.12}$$

hence, by the monotonicity of norm,

$$m\left(\frac{x}{y}\right)\|y\| \leq \|x\| \leq M\left(\frac{x}{y}\right)\|y\| ,$$

i.e.,

$$m\left(\frac{x}{y}\right) \leq 1 \leq M\left(\frac{x}{y}\right) .$$

Now, we prove (ii). For a given $x,y \in \overset{o}{P} \cap S_1$, we have by (1.5.12)

$$\theta \leq x - m\left(\frac{x}{y}\right)y \leq \left[M\left(\frac{x}{y}\right) - m\left(\frac{x}{y}\right)\right]y,$$

which implies

$$\|x - m\left(\frac{x}{y}\right)y\| \leq \left[M\left(\frac{x}{y}\right) - m\left(\frac{x}{y}\right)\right]\|y\| = M\left(\frac{x}{y}\right) - m\left(\frac{x}{y}\right) .$$

It follows therefore

$$\begin{aligned}\|x - y\| &\leq \|x - m\left(\frac{x}{y}\right)y\| + \|m\left(\frac{x}{y}\right)y - y\| \\ &\leq \left[M\left(\frac{x}{y}\right) - m\left(\frac{x}{y}\right)\right] + \left[1 - m\left(\frac{x}{y}\right)\right] \leq 2\left[M\left(\frac{x}{y}\right) - m\left(\frac{x}{y}\right)\right] \\ &= 2\left[\frac{M\left(\frac{x}{y}\right)}{m\left(\frac{x}{y}\right)} - 1\right] m\left(\frac{x}{y}\right) \leq 2\left[\frac{M\left(\frac{x}{y}\right)}{m\left(\frac{x}{y}\right)} - 1\right] = 2\left\{ e^{d(x,y)} - 1\right\}. \square\end{aligned}$$

Lemma 1.5.3. If the norm of E is monotone with respect to P, then $(\overset{o}{P} \cap S_1, d)$ is a complete metric space.

Proof. Suppose the sequence $\{x_n\} \subset \overset{o}{P} \cap S_1$ is a fundamental sequence, i.e., $d(x_n,x_m) \to 0$ $(n,m \to \infty)$. From Lemma 1.5.2 (i), we know

$$m\left(\frac{x_n}{x_m}\right) \to 1 \quad \text{and} \quad M\left(\frac{x_n}{x_m}\right) \to 1 \quad (n,m \to \infty). \tag{1.5.13}$$

For a given $\varepsilon > 0$, choose ε' such that $0 < \varepsilon' < 1$ and

$\ln 1 + \varepsilon'/1 - \varepsilon' < \varepsilon$. On account of (1.5.13), there exists a positive integer N such that

$$1 - \varepsilon' < m\left(\frac{x_n}{x_m}\right) \le 1, \quad 1 \le M\left(\frac{x_n}{x_m}\right) < 1 + \varepsilon', \qquad n > N, \quad m > N.$$

And consequently,

$$(1 - \varepsilon')x_m \le x_n \le (1 + \varepsilon')x_m, \quad \forall\, n > N, \quad m > N. \tag{1.5.14}$$

On the other hand, by virtue of Lemma 1.5.2 (ii), we find $\|x_n - x_m\| \to 0$ $(n,m \to \infty)$, which implies $x_n \to x^* \in E$ and $\|x^*\| = 1$. Now, let m be fixed and $n \to \infty$ in (1.5.14). Then we get

$$(1 - \varepsilon')x_m \le x^* \le (1 + \varepsilon')x_m, \quad \forall\, m > N. \tag{1.5.15}$$

It follows therefore that $x^* \in \overset{o}{P} \quad S_1$ and

$$m\left(\frac{x^*}{x_m}\right) \ge 1 - \varepsilon', \quad M\left(\frac{x^*}{x_m}\right) \le 1 + \varepsilon', \quad \forall\, m > N,$$

which yields

$$d(x^*,x_m) = \ln\left[\frac{M\left(\frac{x^*}{x_m}\right)}{m\left(\frac{x^*}{x_m}\right)}\right] \le \ln \frac{1 + \varepsilon'}{1 - \varepsilon'} < \varepsilon, \quad \forall\, m > N.$$

Thus, $d(x^*,x_m) \to 0$ $(m \to \infty)$ and the completeness of $(\overset{o}{P} \cap S_1,d)$ is proved.

Theorem 1.5.2. If P is normal and solid, then $(\overset{o}{P} \cap S_1,d)$ is a complete metric space, where $S_1 = \{x \in E \mid \|x\| = 1\}$.

Proof. Since P is normal, by Theorem 1.1.1 there exists a norm $\|\cdot\|_1$ of E, which satisfies the following two conditions:

(a) $\|\cdot\|_1$ is equivalent to $\|\cdot\|$, i.e., there exist $\beta > \alpha > 0$ such that $\alpha\|x\| \le \|x\|_1 \le \beta\|x\|$ for any $x \in E$;

(b) Norm $\|\cdot\|_1$ is monotone, i.e., $\theta \le x \le y$ implies $\|x\|_1 \le \|y\|_1$.

By Lemma 1.5.3, $(\overset{o}{P} \cap S_1^{(1)},d)$ is a complete metric space, where

$S_1^{(1)} = \{x \in E \mid \|x\|_1 = 1\}$. Now, we prove that $(\overset{o}{P} \cap S_1, d)$ is a complete metric space too. Let $\{x_n\} \subset \overset{o}{P} \cap S_1$ and $d(x_n, x_m) \to 0$ $(n,m \to \infty)$. Since $\|x_n\| = 1$, we have from (a) that $0 < \alpha \le \|x_n\|_1 \le \beta$ $(n = 1,2,3,\ldots)$. Setting $z_n = x_n/\|x_n\|_1$, we see $z_n \in \overset{o}{P} \cap S_1^{(1)}$ and

$$d(z_n, z_m) = d\left(\frac{x_n}{\|x_n\|_1}, \frac{x_m}{\|x_m\|_1}\right) = d(x_n, x_m) \to 0 \quad (n,m \to \infty).$$

Hence, by the completeness of $(\overset{o}{P} \cap S_1^{(1)}, d)$, there exists $z^* \in \overset{o}{P} \cap S_1^{(1)}$ such that $d(z_n, z^*) \to 0$ $(n \to \infty)$. Since $\|z^*\|_1 = 1$, we have $0 < \beta^{-1} \le \|z^*\| \le \alpha^{-1}$. Now, let $x^* = z^*/\|z^*\|$, then $x^* \in \overset{o}{P} \cap S_1$ and

$$d(x_n, x^*) = d\left(\|x_n\|_1 z_n, \frac{z^*}{\|z^*\|}\right) = d(z_n, z^*) \to 0 \quad (n \to \infty).$$

Hence, $(\overset{o}{P} \cap S_1, d)$ is complete and our theorem is proved. □

<u>Theorem 1.5.3</u>. Let P be normal and solid and $\{x_n\} \subset \overset{o}{P} \cap S_1$, $x \in \overset{o}{P} \cap S_1$, where $S_1 = \{x \in E \mid \|x\| = 1\}$. Then $d(x_n, x) \to 0$ $(n \to \infty)$ if and only if $\|x_n - x\| \to 0$ $(n \to \infty)$.

<u>Proof</u>. Suppose that $d(x_n, x) \to 0$. Then

$$\frac{M\left(\frac{x_n}{x}\right)}{m\left(\frac{x_n}{x}\right)} \to 1 \quad (n \to \infty). \tag{1.5.16}$$

we know

$$m\left(\frac{x_n}{x}\right)x \le x_n \le M\left(\frac{x_n}{x}\right)x,$$

or equivalently,

$$x \le \frac{x_n}{m\left(\frac{x_n}{x}\right)} \le \frac{M\left(\frac{x_n}{x}\right)}{m\left(\frac{x_n}{x}\right)} x. \tag{1.5.17}$$

Since P is normal, it follows from (1.5.16), (1.5.17), and Theorem 1.1.1 that

$$\left\| \frac{x_n}{m\left(\frac{x_n}{x}\right)} - x \right\| \to 0 \quad (n \to \infty), \tag{1.5.18}$$

and hence

$$\frac{\|x_n\|}{m\left(\frac{x_n}{x}\right)} \to \|x\| \quad (n \to \infty) \ . \tag{1.5.19}$$

Observing $\|x_n\| = \|x\| = 1$, (1.5.19) reduces to

$$m\left(\frac{x_n}{x}\right) \to 1 \quad (n \to \infty) \ . \tag{1.5.20}$$

Now, from (1.5.18) and (1.5.20), we get

$$\|x_n - x\| \le \left\|x_n - \frac{x_n}{m\left(\frac{x_n}{x}\right)}\right\| + \left\|\frac{x_n}{m\left(\frac{x_n}{x}\right)} - x\right\|$$

$$= \left|1 - \left[m\left(\frac{x_n}{x}\right)\right]^{-1}\right| + \left\|\frac{x_n}{m\left(\frac{x_n}{x}\right)} - x\right\| \to 0 \quad (n \to \infty) \ ,$$

Thus, we have proved that $d(x_n, x) \to 0$ implies $\|x_n - x\| \to 0$.

Now, we prove the converse conclusion. Suppose $\|x_n - x\| \to 0$. Choose $e \in \overset{o}{P}$. By Theorem 1.5.1, we have $E_e = E$ and the e-norm $\|\cdot\|_e$ is equivalent to the original norm $\|\cdot\|$ and hence

$$\varepsilon_n = \|x_n - x\|_e \to 0 \quad (n \to \infty) \tag{1.5.21}$$

and

$$-\varepsilon_n e \le x_n - x \le \varepsilon_n e \ . \tag{1.5.22}$$

Since $x \in \overset{o}{P}$, we can choose a small positive number r such that

$$x \ge re. \tag{1.5.23}$$

It follows from (1.5.22) and (1.5.23) that

$$\left(1 - \frac{\varepsilon_n}{r}\right)x \le x - \varepsilon_n e \le x_n \le x + \varepsilon_n e \le \left(1 + \frac{\varepsilon_n}{r}\right)x ,$$

which implies

$$1 - \frac{\varepsilon_n}{r} \le m\left(\frac{x_n}{x}\right) \le M\left(\frac{x_n}{x}\right) \le 1 + \frac{\varepsilon_n}{r} ,$$

and therefore

$$d(x_n,x) = \ln\left\{\frac{M\left(\frac{x_n}{x}\right)}{m\left(\frac{x_n}{x}\right)}\right\} \le \ln \frac{r + \varepsilon_n}{r - \varepsilon_n} \to 0 \quad (n \to \infty).$$

Thus, we have proved that $\|x_n - x\| \to 0$ implies $d(x_n,x) \to 0$. □

Theorem 1.5.3 shows that the convergence in Hilbert's projective metric and the convergence in norm are equivalent on $\overset{o}{P} \cap S_1$.

Example 1.5.3. Let P_1 be the solid cone in Example 1.1.1. Then $\overset{o}{P}_1 = \{x = (x_1,x_2,\ldots,x_n) \in R^n \mid x_i > 0, \quad i = 1,2,\ldots,n\}$. For $x = (x_1,x_2,\ldots,x_n) \in \overset{o}{P}_1$ and $y = (y_1,y_2,\ldots,y_n) \in \overset{o}{P}_1$, we have

$$m\left(\frac{x}{y}\right) = \min_i\left(\frac{x_i}{y_i}\right) \quad\text{and}\quad M\left(\frac{x}{y}\right) = \max_i\left(\frac{x_i}{y_i}\right)$$

hence

$$d(x,y\ ) = \ln\left\{\frac{\max_i\left(\frac{x_i}{y_i}\right)}{\min_i\left(\frac{x_i}{y_i}\right)}\right\} = \ln\left(\max_{i,j} \frac{x_i y_j}{x_j y_i}\right) .$$

Since P_1 is normal, we know from Theorem 1.5.2 and Theorem 1.5.3 that $(\overset{o}{P}_1 \cap S_1,d)$ is a complete metric space and $d(x_n,x) \to 0$ is equivalent to $\|x_n - x\| \to 0$ on $\overset{o}{P}_1 \cap S_1$, where $S_1 = \left\{x \in R^n \mid \|x\| = \sqrt{\sum_{i=1}^n x_i^2} = 1\right\}$.

Similarly, if P_2 is the solid cone in Example 1.1.2, then $\overset{o}{P}_2 = \{x(t) \in C(G) \mid x(t) > 0, \quad \forall\ t \in G\}$,

$$d(x,y) = \ln\left\{\frac{\max\limits_{t\in G}\frac{x(t)}{y(t)}}{\min\limits_{t\in G}\frac{x(t)}{y(t)}}\right\} = \ln\left\{\max_{t,x\in G}\frac{x(t)y(s)}{x(s)y(t)}\right\}$$

and $(\overset{o}{P}_2 \cap S_1, d)$ is a complete metric space, where $S_1 = \{x(t) \in C(G) \mid \|x\| = \max\limits_{t\in G}|x(t)| = 1\}$. Moreover, on $\overset{o}{P}_2 \cap S_1$, $d(x_n,x) \to 0$ if and only if $x_n(t)$ converges to $x(t)$ uniformly on $t \in G$.

Example 1.5.4. Let E_e be the real Banach space with norm $\|\cdot\|_e$ given by (1.5.6) and (1.5.7). Let $(P_{10})_e$ be the normal solid cone of E_e given by (1.5.8). Then, for $x(t), y(t) \in (\overset{o}{P}_{10})_e$, we have

$$m\left(\frac{x}{y}\right) = \inf_{0<t<1}\frac{x(t)}{y(t)} \quad\text{and}\quad M\left(\frac{x}{y}\right) = \sup_{0<t<1}\frac{x(t)}{y(t)} .$$

Hence

$$d(x,y) = \ln\left\{\frac{\left(\sup\limits_{0<t<1}\frac{x(t)}{y(t)}\right)}{\left(\inf\limits_{0<t<1}\frac{x(t)}{y(t)}\right)}\right\} = \ln\left(\sup_{0<t,s<1}\frac{x(t)y(s)}{x(s)y(t)}\right) .$$

By Theorems 1.5.2 and 1.5.3, $((\overset{o}{P}_{10})_e \cap S_1, d)$ is a complete metric space and $d(x_n,x) \to 0$ if and only if $x_n(t)/t(1 - t)$ converges to $x(t)/t(1 - t)$ uniformly on $0 < t < 1$ when $\{x_n\} \subset (\overset{o}{P}_{10})_e \cap S_1$ and $x \in (\overset{o}{P}_{10})_e \cap S_1$, where

$$S_1 = \left\{x(t) \in E_e \mid \|x\|_e = \sup_{0<t<1}\frac{2|x(t)|}{t(1 - t)} = 1\right\} .$$

1.6 *Notes and Comments.*

The main results of Theorem 1.1.1 can be found in Peressini [1], Krasnosel'skii [1], and Guo [1] and some parts of its proof are new. Most of the results of Section 1.2, 1.3, and 1.4 are taken from Krasnosel'skii [1], Krein and Rutman [1], and Deimling [1], but Theorem 1.2.2 is due to Sun [1] and Theorem 1.2.3 is proved by Guo. Theorem 1.5.1 is standard and can be found in Krasnosel'skii [1]. Theorem 1.5.2 and Theorem 1.5.3 are from Guo and are not yet published. For related results, see Schaefer [1], Krasnosel'skii and Zabrieko [1], Guo [2], and Bushell [1], [2].

CHAPTER 2. POSITIVE FIXED POINT THEORY

2.0 Introduction

In this chapter, we use the basic properties of cones coupled with the fixed point index to investigate the positive fixed points of some classes of nonlinear operators.

Section 2.1 deals with fixed points of increasing and decreasing operators. We give several theorems about monotone operators from an order interval $[u_0, v_0]$ into itself. Theorems of this type are useful for nonlinear differential and integral equations. On account of the requirement of differential equations, we also introduce the coupled quasi-fixed point theorems of mixed monotone operators from the topological product $[u_0, v_0] \times [u_0, v_0]$ into $[u_0, v_0]$.

In Section 2.2, fixed points of concave operators and convex operators are discussed. We first introduce the definitions of e-concave operators, e-convex operators $(e > \theta)$, strongly sublinear operators, strongly superlinear operators, α-concave operators, and $(-\alpha)$-convex operators $(0 < \alpha < 1)$. We then discuss their fixed points and relative properties. Because investigation of e-convex operators (inparticular, strongly superlinear operators) is somewhat difficult, there are some open problems.

Section 2.3 is devoted to the study of fixed point theorems of cone expansion and compression. We first introduce the notion of fixed point index by means of the Leray-Schauder theory of topological degree. We then use the fixed point index to establish the fixed point theorem of cone expansion and compression, and discuss some of its extensions. We also discuss the fixed points of differential operators.

Section 2.4 deals with multiple fixed point theorems. We establish some results about the existence of two or three fixed points of nonlinear operators.

Finally, Section 2.5 gives a fixed point theorem which can be called the fixed point theorem of domain expansion and compression.

In this chapter, we also give some applications of the general results to nonlinear ordinary and partial differential equations.

2.1 *Fixed Points of Monotone Operators.*

(a) Fixed Points of Increasing Operators.

Let P be a cone in real Banach space E and $\leq$ be the partial ordering defined by P. Let D be a subset of E.

Definition 2.1.1. An operator $A : D \to E$ is said to be increasing if $x_1 \leq x_2$ $(x_1, x_2 \in D)$ implies $Ax_1 \leq Ax_2$, strictly increasing if $x_1 < x_2$ $(x_1, x_2 \in D)$ implies $Ax_1 < Ax_2$, and strongly increasing if $x_1 < x_2$ $(x_1, x_2 \in D)$ implies $Ax_1 \ll Ax_2$ in case $\overset{o}{P} \neq \phi$. Similarly, A is said to be decreasing if $x_1 \leq x_2$ $(x_1, x_2 \in D)$ implies $Ax_1 \geq Ax_2$, strictly decreasing if $x_1 < x_2$ $(x_1, x_2 \in D)$ implies $Ax_1 > Ax_2$, and strongly decreasing if $x_1 < x_2$ $(x_1, x_2 \in D)$ implies $Ax_1 \gg Ax_2$ in case $\overset{o}{P} \neq \phi$.

Definition 2.1.2. An operator $A : D \to E$ is said to be completely continuous if it is continuous and compact. Notice that the compactness means that the set $A(S)$ is relatively compact for any bounded set $S \subset D$. A is called a k-set-contraction $(k \geq 0)$ if it is continuous, bounded and

$$\gamma(A(S)) \leq k\gamma(S) \tag{2.1.1}$$

for any bounded set $S \subset D$, where $\gamma(S)$ denotes the measure of noncompactness of S. Readers will find the definitions and properties of measure of noncompactness in Lakshmikantham and Leela [1] and Guo [1].

A k-set-contraction is called a strict-set-contraction if $k < 1$. An operator A is said to be condensing if it is continuous, bounded, and

$$\gamma(A(S)) < \gamma(S) \tag{2.1.2}$$

for any bounded set $S \subset D$ with $\gamma(S) > 0$.

It is evident that if operator A is completely continuous, then A is a strict-set-contraction, and, if A is a strict-set-contraction, then A is a condensing mapping.

Theorem 2.1.1. Let $u_0, v_0 \in E$, $u_0 < v_0$ and $A : [u_0,v_0] \to E$ be an increasing operator such that

$$u_0 \leq Au_0, \quad Av_0 \leq v_0 \ . \tag{2.1.3}$$

Suppose that one of the following two conditions is satisfied:

(H_1) P is normal and A is condensing;

(H_2) P is regular and A is semicontinuous, i.e., $x_n \to x$ strongly implies $Ax_n \to Ax$ weakly.

Then, A has a maximal fixed point x^* and a minimal fixed point x_* in $[u_0,v_0]$; moreover

$$x_* = \lim_{n\to\infty} v_n, \quad x_* = \lim_{n\to\infty} u_n, \tag{2.1.4}$$

where $v_n = Av_{n-1}$ and $u_n = Au_{n-1}$ $(n = 1,2,3,\ldots)$, and

$$u_0 \leq u_1 \leq \cdots \leq u_n \leq \cdots \leq v_n \leq \cdots \leq v_1 \leq v_0 \ . \tag{2.1.5}$$

Proof. Since A is increasing, it follows from (2.1.3) that (2.1.5) holds. Now, we prove that $\{u_n\}$ converges to some $x_* \in E$ and $Ax_* = x_*$. When (H_1) is satisfied, the set $S = \{u_0,u_1,u_2,\ldots\}$ is bounded and $S = A(S) \cup \{u_0\}$, hence $\gamma(S) = \gamma(A(S))$. Noticing that A is condensing, we have $\gamma(S) = 0$, i.e., S is a relatively compact set. Hence there exists a subsequence $\{u_{n_k}\} \subset \{u_n\}$ such that $u_{n_k} \to x_*$. Clearly, $u_n \leq x_* \leq v_n$ $(n = 1,2,3,\ldots)$. When $m > n_k$, we have $\theta \leq x_* - u_m \leq x_* - u_{n_k}$, and hence, by virtue of normality of P and Theorem 1.1.1, $\|x_* - u_m\| \leq N\|x_* - u_{n_k}\|$, where N denotes the normal constant of P. Thus, $u_m \to x_*$ $(m \to \infty)$. Taking limit $n \to \infty$ on both sides of the equality $u_n = Au_{n-1}$, we get $x_* = Ax_*$ since A is continuous.

When (H_2) is satisfied, $\{u_n\}$ converges to some $x_* \in E$ in view of regularity of P. Since A is semicontinuous, $u_n = Au_{n-1}$ converges to Ax^* weakly and hence $Ax_* = x_*$.

Similarly, we can prove that $\{u_n\}$ converges to some $x^* \in E$ and $Ax^* = x^*$.

Finally, we prove that x^* and x_* are the maximal and minimal fixed points of A in $[u_0,v_0]$, respectively. Let $\bar{x} \in [u_0,v_0]$ and $A\bar{x} = \bar{x}$. Since A is increasing, it follows from $u_0 \leq \bar{x} \leq v_0$ that $Au_0 \leq A\bar{x} \leq Av_0$, i.e., $u_1 \leq \bar{x} \leq v_1$. Using the same argument, we get $u_2 \leq \bar{x} \leq v_2$, and, in general, $u_n \leq \bar{x} \leq v_n$ $(n = 1,2,3,\ldots)$. Now, taking limit $n \to \infty$, we obtain $x_* \leq \bar{x} \leq x^*$, and our theorem is proved. □

<u>Corollary 2.1.1</u>. Let the conditions of Theorem 2.1.1 be satisfied. Suppose that A has only one fixed point $\bar{x}$ in $[u_0,v_0]$. Then, for any $x_0 \in [u_0,v_0]$, the successive iterates

$$x_n = Ax_{n-1} \quad (n = 1,2,3,\ldots) \tag{2.1.6}$$

converge to $\bar{x}$, i.e., $\|x_n - \bar{x}\| \to 0$ $(n \to \infty)$.

<u>Proof</u>. Since $u_0 \leq x_0 \leq v_0$ and A is increasing, we have

$$u_n \leq x_n \leq v_n \quad (n = 1,2,3,\ldots). \tag{2.1.7}$$

By hypotheses we must have $x_* = x^* = \bar{x}$. It follows therefore from (2.1.7) and (2.1.4) normality of P, and Theorem 1.1.1 that $x_n \to \bar{x}$. □

<u>Theorem 2.1.2</u>. Let u_0, $v_0 \in E$, $u_0 < v_0$ and $A : [u_0,v_0] \to E$ be an increasing operator such that (2.1.3) holds. Suppose that P is strongly

minihedral. Then, A has a maximal fixed point x^* and a minimal fixed point x_* in $[u_0,v_0]$.

Proof. Let $D = \{x \in E \mid u_0 \leq x \leq v_0,\ Ax \geq x\}$. Evidently, $u_0 \in D$ and v_0 is an upper bound of D. Consequently, by the strong minihedrality of P, $x^* = \sup D$ exists. Now, we prove that x^* is the maximal fixed point of A in $[u_0,v_0]$. In fact, $u_0 \leq x \leq x^* \leq v_0$ for any $x \in D$ and hence $u_0 \leq Au_0 \leq Ax \leq Ax^* \leq Av_0 \leq v_0$. Therefore, observing $Ax \geq x$, we obtain $x \leq Ax^*$ for any $x \in D$. It follows from the definition of least upper bound that $x^* \leq Ax^*$. On the other hand, from $x^* \leq Ax^*$ we know that $Ax^* \leq A(Ax^*)$, which implies that $Ax^* \in D$, and therefore $Ax^* \leq x^*$. Thus $Ax^* \in x^*$. If $\bar{x}$ is any fixed point of A in $[u_0,v_0]$, then $\bar{x} \in D$, and $\bar{x} \leq x^*$. This proves the maximality of x^*.

Similarly, we can prove that $x_* = \inf D_1$ is the minimal fixed point of A in $[u_0,v_0]$, where $D_1 = \{x \in E \mid u_0 \leq x \leq v_0,\ Ax \leq x\}$. □

Notice that in Theorem 2.1.2 we do not assume A to be continuous. On the other hand, we cannot assert that the limits (2.1.4) exist.

Theorem 2.1.3. Let $u_0,v_0 \in E$, $u_0 < v_0$ and $A = [u_0,v_0] \to E$ be an increasing operator such that (2.1.3) holds. Suppose that $A([u_0,v_0])$ is a relatively compact set of E. Then A has at least one fixed point in $[u_0,v_0]$.

Proof. Let $F = \{x \in A([u_0,v_0]) \mid Ax \geq x\}$. We have $A(Au_0) \geq Au_0$, which implies $Au_0 \in F$, and therefore F is not empty. Since E is partially ordered by P, F is a partially order set. Now, suppose that G is a completely ordered subset of F. Since $A([u_0,v_0])$ is relatively compact by hypothesis and $G \subset F \subset A([u_0,v_0])$, G is relatively compact and so separable, i.e., there exists a denumerable set $V = \{y_1,y_2,\ldots\} \subset G$, which is dense in G. Since G is completely ordered, $z_n = \sup\{y_1,y_2,\ldots,y_n\}$ $(n = 1,2,3,\ldots)$ exist and $z_n \in G$ (in fact, z_n

equals one of y_i, $i = 1,2,\ldots,n$). It follows from the relative compactness of G that there exists a subsequence $\{z_{n_i}\} \subset \{z_n\}$ such that $z_{n_i} \to z^* \in E$. Since

$$z_1 \le z_2 \le \ldots \le z_n \le \ldots ,$$

we have

$$y_n \le z_n \le z^* \quad (n = 1,2,3,\ldots) \tag{2.1.8}$$

and $z^* \in \bar{G} \subset \bar{F} \subset \overline{A([u_0,v_0])} \subset [u_0,v_0]$. From (2.1.8) we know $z \le z^*$ for any $z \in G$, and so $z \le Az \le Az^*$ for any $z \in G$. Thus Az^* is an upper bound of G. On the other hand, from $z_n \le Az^*$ $(n = 1,2,3,\ldots)$ we know $z^* \le Az^*$ and therefore $Az^* \le A(Az^*)$, which shows that $Az^* \in F$. Thus, Az^* is an upper bound of G in F. It follows therefore from Zorn's Lemma that F contains a maximal element x^*. Since $Ax^* \ge x^*$ and so $A(Ax^*) \ge Ax^*$, we have $Ax^* \in F$. By the maximality of x^*, we must have $Ax^* = x^*$. □

Corollary 2.1.2. Let u_0, $v_0 \in E$, $u_0 < v_0$, and $A : [u_0,v_0] \to E$ be an increasing operator such that (2.1.3) holds. Suppose that P is normal and A is compact. Then A has at least one fixed point in $[u_0,v_0]$.

Proof. Since P is normal, $[u_0,v_0]$ is bounded. Thus, the relative compactness of the set $A([u_0,v_0])$ follows from the compactness of operator A. □

Theorem 2.1.4. Let u_0, $v_0 \in E$, $u_0 < v_0$, and $A : [u_0,v_0] \to E$ be an increasing operator such that (2.1.3) holds. Suppose that P is minihedral and $A([u_0,v_0])$ is a relatively compact set of E. Then, A has a maximal fixed point x^* and a minimal fixed point x^* in $[u_0,v_0]$.

Proof. Let $F = \{x \in A([u_0,v_0]) \mid Ax \ge x\}$. By Zorn's Lemma we have already proved in Theorem 2.1.3 that F contains a maximal element x^* and

$Ax^* = x^*$. Now, we prove that x^* is the maximal fixed point of A in $[u_0, v_0]$. In fact, suppose $\bar{x}$ is any fixed point of A in $[u_0, v_0]$, then by the minihedrality of P, $v = \sup\{\bar{x}, x^*\}$ exists. From $v \geq \bar{x}$ and $v \geq x^*$ we get $Av \geq A\bar{x} = \bar{x}$ and $Av \geq Ax^* = x^*$. Hence $v \leq Ax$, and so $Av \leq A(Av)$. It follows therefore $Av \in F$. By the maximality of x^* we have $Av = x^*$, and so $x^* \geq \bar{x}$. Hence, x^* is the maximal fixed point of A in $[u_0, v_0]$.

In the same way we can prove that $F_1 = \{x \in A([u_0, v_0]) \mid Ax \leq x\}$ contains a minimal element x_*, which satisfies $Ax_* = x_*$ and is the minimal fixed point of A in $[u_0, v_0]$. □

Corollary 2.1.3. Let u_0, $v_0 \in E$, $u_0 < v_0$, and $A : [u_0, v_0] \to E$ be an increasing operator such that (2.1.3) holds. Suppose that P is normal and minihedral and A is compact. Then, A has a maximal fixed point x^* and a minimal fixed point x_* in $[u_0, v_0]$.

We remark that Theorems 2.1.3, 2.1.4 and Corollaries 2.1.2, 2.1.3 do not require A to be continuous; thus, they are essentially different from Theorem 2.1.1.

Example 2.1.1. Consider the two-point boundary value problem of ordinary differential equation:

$$-x'' = \lambda f(t,x), \quad x(0) = x(1) = 0, \tag{2.1.9}$$

where λ is a parameter, $f(t,x)$ is defined and continuous on $0 \leq t \leq 1$ and $x \geq 0$, and $f(t,0) \equiv 0$. It is obvious that $x_\lambda(t) \equiv 0$ is a trivial solution of problem (2.1.9) for any λ. Now, suppose (i) $f(t,x)$ is increasing with respect to x, i.e., $0 \leq t \leq 1$, $0 \leq x_1 \leq x_2$ implies $f(t,x_1) \leq f(t,x_2)$, (ii) $f(t,x) > 0$, $\forall\ 0 < t < 1$, $x > 0$ and (iii) $f(t,x)/x$ tends to zero uniformly with respect to $t \in [0,1]$ as $x \to +\infty$. Then, for any $M > 0$ there exists an $R > 0$ such that when $\lambda \geq R$ problem (2.1.9) has at least one nontrivial solution $x_\lambda(t)$, which belongs to $C^2[0,1]$ and satisfies

$$x_\lambda(t) \geq Mt(1 - t), \quad \forall\ 0 \leq t \leq 1. \tag{2.1.10}$$

Proof. It is well known (see, for example, Bernfeld and Lakshmikantham [1]) that the solution (in $C^2[0,1]$) of problem (2.1.9) is equivalent to the solution (in $C[0,1]$) of the integral equation

$$x(t) = \lambda \int_0^1 G(t,s)f(s,x(s))\, ds, \tag{2.1.11}$$

where $G(t,s)$ is the Green function of differential operator $-x''$ with boundary condition $x(0) = x(1) = 0$, given by

$$G(t,s) = \begin{cases} t(1 - s), & t \leq s; \\ s(1 - t), & t > s. \end{cases}$$

Define

$$Ax(t) = \int_0^1 G(t,s)f(s,x(t))\, ds$$

and let $E = C[0,1]$, $P = \{x(t) \in C[0,1] \mid x(t) \geq 0,\ \forall\ 0 \leq t \leq 1\}$. Clearly, P is a normal cone of E and it is easy to show that operator $A : P \to E$ is completely continuous and increasing. Let $u_0(t) = Mt(1 - t)$, where $M > 0$ is given, and

$$w(t) = Au_0(t) = \int_0^t s(1 - t)f(s,u_0(s))\, ds + \int_t^1 t(1 - s)f(s,u_0(s))\, ds.$$

It is easy to calculate

$$w'(t) = \int_0^t (-s)f(s,u_0(s))\, ds + \int_t^1 (1 - s)f(s,u_0(s))\, ds, \quad \forall\ t \in [0,1],$$

and hence

$$w'(0) = \int_0^1 (1 - s)f(s,u_0(s))\, ds > 0, \quad w'(1) = -\int_0^1 sf(s,u_0(s))\, ds < 0. \tag{2.1.12}$$

From (2.1.12) and $u_0(0) = u_0(1) = 0$, it is not difficult to see that there exist sufficiently small positive numbers α and δ such that

$$w(t) \geq \alpha u_0(t), \quad \forall\, t \in [0,\delta] \cup [1-\delta,1]. \tag{2.1.13}$$

On the other hand, it is obvious

$$\min_{\delta \leq t \leq 1-\delta} w(t) > 0. \tag{2.1.14}$$

It follows from (2.1.13) and (2.1.14) that there exists $\gamma > 0$ such that

$$Au_0(t) = w(t) \geq \gamma u_0(t), \quad \forall\, t \in [0,1]. \tag{2.1.15}$$

Now, we set $R = \gamma^{-1}$ and show that this is required R. In fact, for any $\lambda \geq R$, we have, by (2.1.15),

$$\lambda A u_0(t) \geq u_0(t), \quad \forall\, t \in [0,1]. \tag{2.1.16}$$

On the other hand, by virtue of hypothesis (iii), we can choose a positive number c such that $c > M$ and

$$\frac{f(t,c)}{c} \leq \frac{8}{\lambda}, \quad \forall\, t \in [0,1]. \tag{2.1.17}$$

Let $v_0(t) \equiv c$, then $u_0(t) < v_0(t)$, $\forall\, t \in [0,1]$ and

$$\begin{aligned} \lambda A v_0(t) &= \lambda \int_0^1 G(t,s) f(s,c)\, ds \leq 8c \int_0^1 G(t,s)\, ds \\ &= 4ct(1-t) \leq c = v_0(t), \quad \forall\, t \in [0,1]. \end{aligned} \tag{2.1.18}$$

Observing (2.1.16) and (2.1.18) and applying Theorem 2.1.1, we assert that operator λA has at least one fixed point x_λ in $[u_0, v_0]$, i.e.,

$$u_0(t) \leq x_\lambda(t) \leq v_0(t), \quad \forall\, t \in [0,1],$$

and $x_\lambda(t)$ is a solution of equation (2.1.11). □

(b) Fixed Points of Decreasing Operators.

Theorem 2.1.5. Suppose that (i) P is normal and operator $A : P \to P$ is decreasing and condensing; (ii) $A\theta > \theta$ and $A^2\theta > \varepsilon_0 A\theta$, where $\varepsilon_0 > 0$

and θ denotes the zero element of E; and (iii) for any $x \geq \alpha A\theta$ $(\alpha = \alpha(x) > 0)$ and $0 < t < 1$, there exists an $\eta = \eta(x,t) > 0$ such that

$$A(tx) \leq [t(1 + \eta)]^{-1}Ax. \tag{2.1.19}$$

Then, A has exactly one positive fixed point $x^* > \theta$. Moreover, constructing successively sequence $x_n = Ax_{n-1}$ $(n = 1,2,3,\ldots)$ for any initial $x_0 \in P$, we have

$$\|x_n - x^*\| \to 0 \quad (n \to \infty). \tag{2.1.20}$$

Proof. Letting

$$u_0 = \theta, \quad u_n = Au_{n-1} \quad (n = 1,2,3,\ldots), \tag{2.1.21}$$

and observing the decreasing property of A, we can show easily that

$$\theta = u_0 \leq u_2 \leq \cdots \leq u_2 \leq \cdots \leq u_{2n+1} \leq \cdots \leq u_3 \leq u_1 = A\theta, \tag{2.1.22}$$

$$u_{2n} = A^2u_{2n-1}, \quad u_{2n+1} = A^2u_{2n-1} \quad (n = 1,2,3,\ldots), \tag{2.1.23}$$

and

$$u_{2n} = Au_{2n-1}, \quad u_{2n+1} = Au_{2n} \quad (n = 1,2,3,\ldots). \tag{2.1.24}$$

Since $A^2 : P \to P$ is increasing and condensing and $u_0 \leq A^2u_0$, $A^2u_1 \leq u_1$, it follows from Theorem 1.6.1 that $u_{2n} \to z_*$, $u_{2n+1} \to z^*$ $(n \to \infty)$, $A^2z_* = z_*$, $A^2z^* = z^*$ and z_* and z^* are the minimal and maximal fixed points of A^2 in $[u_0,u_1]$, respectively. Taking limit in (2.1.24), we find

$$z_* = Az^*, \quad \text{and} \quad z^* = Az_*. \tag{2.1.25}$$

Evidently

$$\theta < \varepsilon_0 A\theta \le A^2\theta = u_2 \le u_{2n} \le z_* \le z^* \le u_{2n+1} \quad (n = 1,2,3,\ldots), \quad (2.1.26)$$

and hence

$$z_* \ge \varepsilon_0 A\theta = \varepsilon_0 u_1 \ge \varepsilon_0 z^* .$$

Let $t_0 = \sup\{t > 0 \mid z_* \ge tz^*\}$, so that $0 < \varepsilon_0 \le t_0 < +\infty$ and $z_* \ge t_0 z^*$. Moreover, we must have $t_0 \le 1$, since $z_* \le z^*$. We now prove $t_0 = 1$. If otherwise $0 < t_0 < 1$. By hypothesis (iii), there exists an $\eta_0 > 0$ such that

$$z^* = Az_* \le A(t_0 z^*) \le [t_0(1 + \eta_0)]^{-1} Az^* = [t_0(1 + \eta)]^{-1} z_* ,$$

and hence $z^* \ge t_0(1 + \eta_0)z^*$, which contradicts the definition of t_0.

Thus $t_0 = 1$ and $z_* \ge z^*$, and therefore

$$z_* = z^*. \quad (2.1.27)$$

From (2.1.25) and (2.1.27), we see $Az^* = z^*$, i.e., z^* is a positive fixed point of A.

Finally, we prove that (2.1.20) holds for any positive fixed point x^* of A and any initial $x_0 \in P$, which at the same time implies the uniqueness of positive fixed points of A. Since $x_0 \ge \theta$, we have $\theta \le Ax_0 \le A\theta$, i.e., $u_0 \le x_1 \le u_1$; operating this inequality by A, we find $u_2 \le x_2 \le u_1$. Continuing this process, we obtain in general

$$u_{2n} \le x_{2n} \le u_{2n-1}, \; u_{2n} \le x_{2n+1} \le u_{2n-1}, \; (n = 1,2,3,\ldots). \quad (2.1.28)$$

In the same way, we get

$$u_{2n} \le x^* \le u_{2n-1} \quad (n = 1,2,3,\ldots). \quad (2.1.29)$$

Since $u_{2n} \quad z_* = z^*$, $u_{2n-1} \to z^*$ and P is normal, it follows from

(2.1.28) and Theorem 1.1.1 that $x_{2n} \to z^*$ and $x_{2n+1} \to z^*$. Thus $\|x_n - z^*\| \to 0$ $(n \to \infty)$. On the other hand, taking limit in (2.1.29), we get $x^* = z^*$, and hence $\|x_n - x^*\| \to 0$. The proof is therefore complete. □

Corollary 2.1.4. When P is solid, Theorem 2.1.5 is still true if we replace the hypothesis (iii) by the following condition: (iii*) for any $x \geq \alpha A\theta$ $(\alpha = \alpha(x) > 0)$ and $0 < t < 1$, we have

$$A(tx) << t^{-1}Ax. \tag{2.1.30}$$

Proof. Since $t^{-1}Ax - A(tx) \in \overset{o}{P}$, there exists sufficiently small $(0 < \varepsilon < 1)$ such that

$$t^{-1}Ax - A(tx) - \varepsilon t^{-1}Ax \geq \theta,$$

which implies

$$A(tx) \leq t^{-1}(1 - \varepsilon)Ax = [t(1 + \eta)]^{-1}Ax,$$

where $\eta = \varepsilon/(1 - \varepsilon) > 0$. Thus, (iii*) implies (iii). □

It is easy to see that hypothesis (iii) of Theorem 2.1.5 is equivalent to the following condition: for any $x \geq \alpha A\theta$ $(\alpha = \alpha(x) > 0)$ and $t > 1$, there exist an $\eta = \eta(x,t)$, $0 < \eta < 1$, such that

$$A(tx) \geq [t(1 - \eta)]^{-1}Ax. \tag{2.1.31}$$

Similarly, condition (iii*) is equivalent to: for any $x \geq \alpha A\theta$ $(\alpha = \alpha(x) > 0)$ and $t > 1$, we have

$$A(tx) >> t^{-1}Ax. \tag{2.1.32}$$

Theorem 2.1.6. Suppose that (i) P is normal and solid and operator $A : P \to P$ is strongly decreasing and condensing, (ii) $A\theta > \theta$ and $A^2\theta \geq \varepsilon_0 A\theta$, where $\varepsilon_0 > 0$ and (iii) for any $x \geq \alpha A\theta$ $(\alpha = \alpha(x) > 0)$ and $0 < t < 1$, we have

$$A(tx) < t^{-1}Ax. \tag{2.1.33}$$

Then, A has exactly one positive fixed point $x^* > \theta$ and (2.1.20) holds.

Proof. The proof is the same as that of Theorem 2.1.5. The only difference is how to establish (2.1.27). In the present case, we can proceed as follows. As before, we let $t_0 = \sup\{t > 0 \mid z_* \geq tz^*\}$, and we need to prove $t_0 = 1$. If otherwise $0 < t_0 < 1$, we have by hypothesis (iii),

$$z^* = Az_* \leq A(t_0 z^*) < t_0^{-1} Az^* = t_0^{-1} z_* .$$

Since A is strongly decreasing, it follows from $z_* > t_0 z^*$ that

$$z^* = Az_* << A(t_0 z^*) < t_0^{-1} z_* ,$$

and hence, there exists sufficiently small $\delta_0 > 0$ such that

$$t_0^{-1} z_* - z^* - \delta_0 z^* \geq \theta ,$$

i.e.,

$$z_* \geq t_0(1 + \delta_0) z^* ,$$

which contradicts the definition of t_0.

Notice, hypothesis (iii) of Theorem 2.1.6) is equivalent to: for any $x \geq \alpha A\theta$ $(\alpha = \alpha(x) > 0)$ and $t > 1$, we have

$$A(tx) > t^{-1}Ax . \tag{2.1.34}$$

Example 2.1.2. Let Ω be a bounded convex domain in R^n $(n \geq 2)$ whose boundary $\partial\Omega$ belongs to $C^{2+\mu}$ for some $0 < \mu < 1$. Consider a uniformly elliptic differential operator on $\overline{\Omega}$:

$$Lu = - \sum_{i,j=1}^{n} a_{ij}(x) \frac{\partial^2 u}{\partial x_i \partial x_j} + \sum_{i=1}^{n} b_i(x) \frac{\partial u}{\partial x_i} + c(x)u ,$$

i.e., there exists a positive constant μ_0 such that

$$\sum_{i,j=1}^{n} a_{ij}(x)\xi_i\xi_j \geq \mu_0|\xi|^2$$

for any $x \in \bar{\Omega}$ and $\xi = (\xi_1,\xi_2,\ldots,\xi_n) \in R^n$, and $a_{ij}(x) = a_{ij}(x)$, $c(x) \geq 0$. Here all functions $a_{ij}(x)$, $b_i(x)$, and $c(x)$ belong to $C^{\mu}(\bar{\Omega})$.

Now, we consider the Dirichlet problem:

$$Lu = [a_0(x) + \sum_{i=1}^{m} a_i(x)u^{\alpha_i}]^{-1}, \quad u|_{\partial\Omega} = 0. \tag{2.1.35}$$

Suppose that $a_i(x) \in C^{\mu}(\bar{\Omega})$ $(i = 0,1,2,\ldots,m)$, $a_0(x)$ is positive, and $a_i(x)$ $(i = 1,2,\ldots,m)$ are all nonnegative for any $x \in \bar{\Omega}$. Suppose, in addition, that $0 < \alpha_1 < \alpha_2 < \ldots < \alpha_m \leq 1$ and $m \geq 1$. Then, the Dirichlet problem (2.1.35) has exactly one positive solution $u^*(x)$ on Ω, which belongs to $C^{2+\alpha}(\bar{\Omega})$, where $\alpha = \min\{\mu,\alpha_1,\alpha_2,\ldots,\alpha_m\}$. Moreover, for any initial function $u_0(x)$ which is nonnegative and belongs to $C^{\mu}(\bar{\Omega})$, denoting the unique solution of the linear Dirichlet problem

$$\begin{cases} Lu = \{a_0(x) + \sum_{i=1}^{m} a_i(x)[u_{k-1}(x)]^{\alpha_i}\}^{-1}, \\ u|_{\partial\Omega} = 0 \end{cases} \tag{2.1.36}$$

by $u_k(x)$ $(u_k(x) \in C^{2+\alpha}(\bar{\Omega}))$, $k = 1,2,3,\ldots$, the sequence $\{u_k(x)\}$ must converge to $u^*(x)$ uniformly on $\bar{\Omega}$.

Proof. It is well known that the solution of the Dirichlet problem (2.1.35) is equivalent to the fixed point of the integral operator A:

$$Au(x) = \int_{\bar{\Omega}} G(x,y)\{a_0(y) + \sum_{i=1}^{m} a_i(y)[u(y)]^{\alpha_i}\}^{-1}dy ,$$

where $G(x,y)$ denotes the Green function of differential operator L with boundary condition $u|_{\partial\Omega} = 0$. It is also well known that $G(x,y)$ satisfies the following inequality:

$$0 < G(x,y) < \begin{cases} K_0 \ |x - y|^{2-n}, & n > 2 \\ K_0 \ |\ln|x-y|| \ , & n = 2 \end{cases} \qquad (x, y \in \Omega, x \neq y), \tag{2.1.37}$$

Hence (see Guo [4]), the linear integral operator

$$Gv(x) = \int_{\overline{\Omega}} G(x,y)v(y)dy \tag{2.1.38}$$

is a completely continuous operator from $C(\overline{\Omega})$ into $C(\overline{\Omega})$, and therefore, operator A maps P into P and is completely continuous, where $P = \{u(x) \in C(\overline{\Omega}) \mid u(x) \geq 0, \ \forall \ x \in \overline{\Omega}\}$ is a normal cone of space $C(\overline{\Omega})$. Evidently, A is decreasing, and so A satisfies condition (i) of Theorem 2.1.5. The verification of condition (ii) is trivial. We shall verify condition (iii). For any $u(x) \in P$ and $0 < t < 1$, we have

$$A(tu(x)) \leq t^{-1} \int_{\overline{\Omega}} G(x,y)\{t^{-1}a_0(y) + \sum_{i=1}^{n} a_i(y)[u(y)]^{\alpha_i}\}^{-1} dy. \tag{2.1.39}$$

On the other hand, since $a_0(y) > 0$ for any $y \in \overline{\Omega}$, we find

$$t^{-1}a_0(y) + \sum_{i=1}^{m} a_i(y)[u(y)]^{\alpha_i} > a_0(y) + \sum_{i=1}^{m} a_i(y)[u(y)]^{\alpha_i}, \quad \forall \ y \in \overline{\Omega}.$$

Consequently,

$$\min_{y\in\Omega}\{t^{-1}a_0(y) + \sum_{i=1}^{m} a_i(y)[u(y)]^{\alpha_i}\}/\{a_0(y) + \sum_{i=1}^{m} a_i(y)[u(y)]^{\alpha_i}\}$$

$$= 1 + \eta, \tag{2.1.40}$$

where $\eta > 0$. It follows from (2.1.39) and (2.1.40) that

$$A(tu(x)) \leq [t(1 + \eta)]^{-1} \int_{\overline{\Omega}} G(x,y)\{a_0(y) + \sum_{i=1}^{m} a_i(y)[u(y)]^{\alpha_i}\}^{-1} dy$$

$$= [t(1 + \eta)]^{-1} Au(x),$$

i.e., A satisfies condition (iii) of Theorem 2.1.5. Thus, by Theorem 2.1.5, A has exactly one positive fixed point $u^*(x)$ $(u^*(x) \in P$, $u^*(x) \not\equiv 0)$ and for any $u_0(x) \in P \cap C^1(\overline{\Omega})$, the sequence $\{u_k(x)\}$ converges to $u^*(x)$ uniformly on $\overline{\Omega}$, where

$$u_k(x) = \int_\Omega G(x,y)\{a_0(y) + \sum_{i=1}^m a_i(y)[u_{k-1}(y)]^{\alpha_i}\}^{-1} dy \quad (k = 1,2,3,\ldots). \tag{2.1.41}$$

Now we prove $u^*(x) \in C^{2+\alpha}(\bar{\Omega})$. Since $\{a_0(x) + \sum_{i=1}^m a_i(x)[u^*(x)]^{\alpha_i}\}^{-1}$ is continuous on $\bar{\Omega}$, $u^*(x) = Au^*(x)$ belongs to $C^{1+\delta}(\bar{\Omega})$, where $0 < \delta < 1$ (see Amann [1]), hence $[u^*(x)]^{\alpha_i} \in C^{\alpha_i}(\bar{\Omega})$ $i = 1,2,\ldots,m$ (see Guo [5]), and so it is easy to see that $\{a_0(x) + \sum_{i=1}^m a_i(x)[u^*(x)]^{\alpha_i}\}^{-1}$ belongs to $C^\alpha(\bar{\Omega})$, where $\alpha = \min\{\mu,\alpha_1,\ldots,\alpha_m\}$. Hence, by a classical result (see Gilbarg and Trudinger [1]), we know $u^*(x) = Au^*(x) \in C^{2+\alpha}(\bar{\Omega})$.

Similarly, from (2.1.41) we can deduce $u_k(x) \in C^{2+\alpha}(\bar{\Omega})$, $k = 1,2,3,\ldots$, and (2.1.41) is equivalent to:

$$\begin{cases} Lu_k(x) = \{a_0(x) + \sum_{i=1}^m a_i(x)[u_{k-1}(x)]^{\alpha_i}\}^{-1}, \\ u_k(x)|_{\partial\Omega} = 0 \end{cases} \quad (k = 1,2,3,\ldots),$$

i.e., $u_k(x)$ is the unique solution of the linear Dirichlet problem (2.1.36). □

(c) Coupled Quasi-fixed Points of Mixed Monotone Operators.

In order to investigate monotone iterative techniques for nonlinear differential equations (see Ladde, Lakshmikantham, and Vatsala [1]), we introduce the following notions.

Definition 2.1.3. Let D be a subset of E. Let operator $A : D \times D \to E$. A is said to be mixed monotone if $A(x,y)$ is increasing in x and decreasing in y, i.e., if $x_1 \le x_2$ $(x_1,x_2 \in D)$ implies $A(x_1,y) \le A(x_2,y)$ for any $y \in D$ and $y_1 \le y_2$ $(y_1,y_2 \in D)$ implies $A(x,y_1) \ge A(x,y_2)$ for any $x \in D$. Point $(x^*,y^*) \in D \times D$ is said to be a coupled quasi-fixed point of A if $A(x^*,y^*) = x^*$ and $A(y^*,x^*) = y^*$. $x^* \in D$ is called a fixed point of A if $A(x^*,x^*) = x^*$.

Evidently, if x^* is a fixed point of A, then (x^*,x^*) is a coupled quasi-fixed point of A.

Theorem 2.1.7. Let $u_0, v_0 \in E$, $u_0 < v_0$ and $A : [u_0,v_0] \times [u_0,v_0] \to E$ be a mixed monotone operator such that

$$u_0 \le A(u_0,v_0), \quad A(v_0,u_0) \le v_0 \ . \tag{2.1.42}$$

Suppose that one of the following two conditions is satisfied:

(H_1) P is normal and A is completely continuous;

(H_2) P is regular and A is demicontinuous, i.e., $x_n \to x$ and $y_n \to y$ strongly implies $A(x_n,y_n) \to A(x,y)$ weakly.

Then, A has a coupled quasi-fixed point $(x^*,y^*) \in [u_0,v_0] \times [u_0,v_0]$, which is minimal and maximal in the sense that $x^* \le \bar{x} \le y^*$ and $x^* \le \bar{y} \le y^*$ for any coupled quasi-fixed point $(\bar{x},\bar{y}) \in [u_0,v_0] \times [u_0,v_0]$ of A; moreover, we have

$$x^* = \lim_{n\to\infty} u_n, \quad y^* = \lim_{n\to\infty} v_n, \tag{2.1.43}$$

where $u_n = A(u_{n-1},v_{n-1})$ and $v_n = A(v_{n-1},u_{n-1})$ $(n = 1,2,3,\ldots)$, which satisfy

$$u_0 \le u_1 \le \cdots \le u_n \le \cdots \le v_n \le \cdots \le v_1 \le v_0 \ . \tag{2.1.44}$$

Proof. From (2.1.42) we know $u_0 \le u_1 \le v_1 \le v_0$. Suppose $u_{n-1} \le u_n \le v_n \le v_{n-1}$. Since A is mixed monotone,

$$u_n = A(u_{n-1},v_{n-1}) \le A(u_n,v_{n-1}) \le A(u_n,v_n) = u_{n+1},$$

$$v_n = A(v_{n-1},u_{n-1}) \ge A(v_n,u_{n-1}) \ge A(v_n,u_n) = v_{n+1},$$

and $u_{n+1} = A(u_n,v_n) \le A(v_n,u_{n-1}) \le A(v_n,u_n) = v_{n+1}$.

Hence, by induction, (2.1.44) holds. When (H_1) is satisfied, the set $S = \{u_1, u_2, u_3, \ldots\}$ is relatively compact by virtue of the complete continuity of A, and hence, from the proof of Theorem 2.1.1, we know $u_n \to x^* \in [u_0, v_0]$. When (H_2) is satisfied, $u_n \to x^* \in E$ is deduced directly from the regularity of P. Similarly, we can prove that $\{u_n\}$ converges to some $y^* \in [u_0, v_0]$.

Since A is demicontinuous in both cases (H_1) and (H_2), $u_n = A(u_{n-1}, v_{n-1})$ converges to $A(x^*, y^*)$ weakly and $v_n = A(v_{n-1}, u_{n-1})$ converges to $A(y^*, x^*)$ weakly. Thus $A(x^*, y^*) = x^*$ and $A(y^*, x^*) = y^*$, i.e., (x^*, y^*) is a coupled quasi-fixed point of A.

Finally, we prove the minimal and maximal property of (x^*, y^*). Assume $(\bar{x}, \bar{y}) \in [u_0, v_0] \times [u_0, v_0]$ is any coupled quasi-fixed point of A. Since $u_0 \le \bar{x} \le v_0$ and $u_0 \le \bar{y} \le v_0$,

$$u_1 = A(u_0, v_0) \le A(u_0, \bar{y}) \le A(\bar{x}, \bar{y}) = \bar{x} \le A(v_0, \bar{y}) \le A(v_0, u_0) = v_1,$$

and $u_1 = A(u_0, v_0) \le A(u_0, \bar{x}) \le A(\bar{y}, \bar{x}) = \bar{y} \le A(v_0, \bar{x}) \le A(v_0, u_0) = v_1$. Similarly, $u_2 \le \bar{x} \le v_2$, $u_2 \le \bar{y} \le v_2$ and, in general,

$$u_n \le \bar{x} \le v_n, \quad u_n \le \bar{y} \le v_n \quad (n = 0, 1, 2, \ldots). \tag{2.1.45}$$

Taking limit in (2.1.45), we get $x^* \le \bar{x} \le y^*$ and $x^* \le \bar{y} \le y^*$. □

Corollary 2.1.5. Let the conditions of Theorem 2.1.7 be satisfied. Suppose there exists $0 < \alpha < 1$ such that

$$\|A(x,y) - A(y,x)\| \le \alpha \|x - y\|, \quad (x,y) \in [u_0, v_0] \times [u_0, v_0]. \tag{2.1.46}$$

Then, A has exactly one fixed point $\bar{x}$ in $[u_0, v_0]$ and, if we successively construct the sequences

$$\begin{cases} x_n = A(x_{n-1}, y_{n-1}) \\ y_n = A(y_{n-1}, x_{n-1}) \end{cases} \quad (n = 1, 2, 3, \ldots) \tag{2.1.47}$$

for any initial $(x_0,y_0) \in [u_0,v_0] \times [u_0,v_0]$, we have

$$\|x_n - \bar{x}\| \to 0, \quad \|y_n - \bar{x}\| \to 0 \quad (n \to \infty). \tag{2.1.48}$$

Proof. By (2.1.46) we know

$$\|v_n - u_n\| = \|A(v_{n-1},u_{n-1}) - A(u_{n-1},v_{n-1})\| \leq \alpha\|v_{n-1} - u_{n-1}\|$$
$$(n = 1,2,3,\ldots),$$

and so

$$\|v_n - u_n\| \leq \alpha^n\|v_0 - u_0\| \to 0 \quad (n \to \infty).$$

Hence, by (2.1.43), $x^* = y^*$. Let $\bar{x} = x^* = y^*$, then $\bar{x}$ is a fixed point of A. By virtue of the minimal and maximal property (x^*,y^*), we show easily that $\bar{x}$ is the unique fixed point of A in $[u_0,v_0]$.

Now, let $(x_0,y_0) \in [u_0,v_0] \times [u_0,v_0]$ be given and (2.1.47) be constructed. Similar to the establishment of (2.1.45), we get

$$u_n \leq x_n \leq v_n, \quad u_n \leq y_n \leq v_n, \quad (n = 0,1,2,\ldots). \tag{2.1.49}$$

It follows therefore from (2.1.43), (2.1.49), that $\bar{x} = x^* = y^*$ and the normality of P show that (2.1.48) holds. □

Theorem 2.1.8. Let u_0, $v_0 \in E$, $u_0 < v_0$, and $A : [u_0,v_0] \times [u_0,v_0] \to E$ be a mixed monotone operator such that (2.1.42) holds. Suppose that P is strongly minihedral. Then, A has a coupled quasi-fixed point $(x^*,y^*) \in [u_0,v_0] \times [u_0,v_0]$ with $x^* \geq y^*$, which is maximal and minimal in the sense that $y^* \leq \bar{x} \leq x^*$ and $y^* \leq \bar{y} \leq x^*$ for any coupled quasi-fixed point $(\bar{x},\bar{y}) \in [u_0,v_0] \times [u_0,v_0]$ of A.

Proof. Let $D = \{(x,y) \in [u_0,v_0] \times [u_0,v_0] \mid A(x,y) \geq x,\ A(y,x) \leq y\}$, then $(u_0,v_0) \in D$ and hence $D \neq \phi$. Let

$$D_1 = \{x \mid (x,y) \in D \text{ for some } y, \text{ here } y \text{ depends on } x\},$$

and

$$D_2 = \{y \mid (x,y) \in D \text{ for some } x, \text{ here } x \text{ depends on } y\}.$$

Evidently, $D_1 \neq \phi$, $D_2 \neq \phi$, and v_0 and u_0 are upper bound and lower bound of both D_1 and D_2, respectively, and so, by the strong minihedrality of P, $x^* = \sup D_1$ and $y^* = \inf D_2$ exist and $(x^*,y^*) \in [u_0,v_0] \times [u_0,v_0]$.

Now, for any $(x,y) \in D$, we have

$$A(x^*,y^*) \geq A(x,y^*) \geq A(x,y) \geq x,$$

$$A(y^*,x^*) \leq A(y,x^*) \leq A(y,x) \leq y,$$

and hence

$$A(x^*,y^*) \geq x^*, \quad A(y^*,x^*) \leq y^*. \tag{2.1.50}$$

Let $x_* = A(x^*,y^*)$ and $y_* = A(y^*,x^*)$, then

$$u_0 \leq A(u_0,v_0) \leq A(u_0,y^*) \leq A(x^*,y^*) = x_* \leq A(v_0,y^*) \leq A(v_0,u_0) \leq v_0.$$

Similarly, we have $u_0 \leq y_* \leq v_0$. Thus, by (2.1.50)

$$A(x_*,y_*) \geq A(x^*,y_*) \geq A(x^*,y^*) = x_*,$$

$$A(y_*,x_*) \leq A(y^*,x_*) \leq A(y^*,x^*) = y_*.$$

Consequently $(x_*,y_*) \in D$, and therefore

$$x_* \leq x^*, \quad y_* \geq y^*. \tag{2.1.51}$$

It follows from (2.1.50) and (2.1.51) that $x_* = x^*$ and $y_* = y^*$, i.e., $A(x^*,y^*) = x^*$ and $A(y^*,x^*) = y^*$. Thus (x^*,y^*) is a coupled quasi-fixed point of A.

Finally, let $(\bar{x},\bar{y}) \in [u_0,v_0] \times [u_0,v_0]$ be any coupled quasi-fixed point of A, then $(\bar{x},\bar{y}) \in D$ and therefore $\bar{x} \leq x^*$ and $\bar{y} \geq y^*$. Since $(\bar{y},\bar{x})$ is also a coupled quasi-fixed point of A, we have $\bar{y} \leq x^*$, $\bar{x} \geq y^*$. □

Notice, in Theorem 2.1.8 we do not assume A to be continuous. On the other hand, we cannot assert the limits (2.1.43) exist.

2.2 *Fixed Points of Concave and Convex Operators.*

As before, let P be a cone in real Banach space E and $\leq$ be the partial ordering defined by P.

<u>Definition 2.2.1</u>. Let operator $A : P \to P$ and $e > \theta$. Suppose that (i) for any $x > \theta$ there exist $\alpha = \alpha(x) > 0$ and $\beta = \beta(x) > 0$ such that

$$\alpha e \leq Ax \leq \beta e \tag{2.2.1}$$

and (ii) for any $x \in P$ satisfying $\alpha_1 e \leq x \leq \beta_1 e$ $(\alpha_1 = \alpha_1(x) > 0,$ $\beta_1 = \beta_1(x) > 0)$ and any $0 < t < 1$, there exists $\eta = \eta(x,t) > 0$ such that

$$A(tx) \geq (1 + \eta)tAx. \tag{2.2.2}$$

Then, A is called a <u>e-concave operator</u>.

Similarly, if in the above definition we replace condition (ii) by the following condition (ii'): for any $x \in P$ satisfying $\alpha_1 e \leq x \leq \beta_1 e$ $(\alpha_1 = \alpha_1(x) > 0,$ $\beta_1 = \beta_1(x) > 0)$ and any $0 < t < 1$, there exists $\eta = \eta(x,t) > 0$ such that

$$A(tx) \leq (1 - \eta)tAx, \tag{2.2.3}$$

then A is called a <u>e-convex operator</u>.

<u>Definition 2.2.2</u>. Let P be solid and $A : P \to P$. If

$$A(tx) \gg tAx, \quad \forall\ x > \theta, \quad 0 < t < 1, \tag{2.2.4}$$

then A is said to be strongly sublinear. Similarly, if

$$A(tx) \ll tAx, \quad \forall\ x > \theta, \quad 0 < t < 1, \tag{2.2.5}$$

then A is said to be strongly superlinear.

Example 2.2.1. Consider the nonlinear integral operator

$$A\psi(x) = \int_G k(x,y)[\psi(y)]^\alpha dy , \tag{2.2.6}$$

where G is a bounded closed set of Euclidean space R^n and $\alpha > 0$, $\alpha \neq 1$. Suppose that $k(x,y)$ is continuous and positive on $G \times G$, then $A : P \to P$ is completely continuous, where $P = \{\psi(x) \in C(G) \mid \psi(x) \geq 0, x \in G\}$ is a cone of $C(G)$.

(a) if $0 < \alpha < 1$, then, for any $\psi > \theta$ and $0 < t < 1$, we have

$$A[t\psi(x)] - tA[\psi(x)] = (t^\alpha - t) \int_G k(x,y)[\psi(y)]^\alpha dy$$

$$\geq m(t^\alpha - t) \int_G [\psi(y)]^\alpha dy = \beta > 0 ,$$

where $m = \min\limits_{x,y \in G} k(x,y) > 0$. Hence $A(t\psi) - tA\psi \in \overset{o}{P}$, i.e., $A(t\psi) \gg tA\psi$. Thus, A is strongly sublinear.

(b) if $\alpha > 1$, then for any $\psi > \theta$ and $0 < t < 1$, we find

$$tA[\psi(x)] - A[t\psi(x)] = (t - t^\alpha) \int_G k(x,y)[\psi(y)]^\alpha dy$$

$$\geq m(t - t^\alpha) \int_G [\psi(y)]^\alpha dy = \gamma > 0,$$

hence $A(t\psi) \ll tA\psi$, i.e., A is strongly superlinear.

This example shows the source of the names "strongly sublinear operator" and "strongly superlinear operator".

Example 2.2.2. Let L be the uniformly elliptic differential operator of Example 2.1.2 and consider the Dirichlet problem:

$$\begin{cases} Lu = f(x,u), \\ u|_{\partial\Omega} = 0, \end{cases} \tag{2.2.7}$$

where $f(x,u)$ is nonnegative and continuous on $x \in \bar{\Omega}$ and $u \geq 0$. As pointed out in Example 2.1.2, finding the solution of problem (2.2.7) is equivalent to finding the fixed point of the integral operator A:

$$Au(x) = \int_{\bar{\Omega}} G(x,y)f(y,u(y))dy\ , \tag{2.2.8}$$

where $G(x,y)$ is the corresponding Green function, which satisfies (2.1.37), and the linear integral operator (2.1.38) is completely continuous from $C(\bar{\Omega})$ into $C(\bar{\Omega})$. Moreover, we know that operator A is completely continuous from P into P, where $P = \{u(x) \in C(\bar{\Omega}) \mid u(x) \geq 0, \forall\ x \in \bar{\Omega}\}$. Now we prove that if the function $f(x,u)$ satisfies

$$f(x,tu) > tf(x,u), \quad \forall\ u > 0, \quad x \in \Omega, \quad 0 < t < 1, \tag{2.2.9}$$

then operator A is e-concave, where

$$e(x) = \int_{\bar{\Omega}} G(x,u)dy, \quad \forall\ x \in \bar{\Omega}. \tag{2.2.10}$$

Proof. Here, we need to use a conclusion about integral operator (2.1.38), you can find it in Krasnosel'skii [1] or Amann [1]: linear integral operator (2.1.38) is e-positive, i.e., for any $v > \theta$, there exists $\alpha = \alpha(v) > 0$ and $\beta = \beta(v) > 0$ such that $\alpha e \leq Gv \leq \beta e$, i.e.,

$$\alpha \int_{\bar{\Omega}} G(x,u)dy \leq \int_{\bar{\Omega}} G(x,y)v(y)dy \leq \beta \int_{\bar{\Omega}} G(x,y)dy, \quad \forall\ x \in \bar{\Omega}. \tag{2.2.11}$$

Now, let $u > \theta$. Then there exists an $x_0 \in \Omega$ such that $u(x_0) > 0$, and it follows from (2.2.9) that

$$f(x_0,u(x_0)) = f(x_0,\tfrac{1}{2}\cdot 2u(x_0)) > \tfrac{1}{2}f(x_0,2u(x_0)) \geq 0.$$

Consequently $fu > \theta$, where f denotes the Nemitskyi operator:

$$fu(x) = f(x,u(x)) \; . \tag{2.2.12}$$

Thus, from (2.2.11) we know that there exist $\alpha > 0$ and $\beta > 0$ such that

$$\alpha e \leq Au = Gfu \leq \beta e \; , \tag{2.2.13}$$

i.e., A satisfies condition (i) of Definition 2.2.1.

Next, suppose $u \in P$, satisfying $\alpha_1 e \leq u \leq \beta_1 e$ $(\alpha_1 = \alpha_1(u) > 0$, $\beta_1 = \beta_1(u) > 0)$ and $0 < t < 1$. Since $e(x) > 0$ for any $x \in \Omega$, we have by (2.2.9)

$$f(x,tu(x)) - tf(x,u(x)) > 0, \; \forall \; x \in \Omega,$$

and hence, by (2.2.11), there exists $\alpha_2 > 0$ such that

$$\int_{\overline{\Omega}} G(x,y)\{f(y,tu(y)) - tf(y,u(y))\}dy \geq \alpha_2 e(x), \quad \forall \; x \in \overline{\Omega}. \tag{2.2.14}$$

On the other hand, it is evident

$$\int_{\overline{\Omega}} G(x,y)f(y,u(y))dy \leq Me(x), \quad \forall \; x \in \overline{\Omega}, \tag{2.2.15}$$

where

$$M = \max_{x \in \overline{\Omega}} f(x,u(x)) .$$

It follows therefore from (2.2.14) and (2.2.15) that

$$\int_{\overline{\Omega}} G(x,y)f(y,tu(y))dy \geq t\left(1 + \frac{\alpha_2}{Mt}\right) \int_{\overline{\Omega}} G(x,y)f(y,u(y))dy, \quad \forall \; x \in \overline{\Omega},$$

i.e., $A(tu) \geq t(1 + \eta)Au$, where $\eta = \alpha_2/Mt > 0$. Thus, A satisfies condition (ii) of Definition 2.2.1, and therefore, A is e-concave.□

Theorem 2.2.1. Let P be solid, $e \in \overset{o}{P}$ and operator $A : P \to P$. Then the following conclusions hold:

(a) if A is strongly sublinear, then A is e-concave;

(b) if A is strongly superlinear, then A is e-convex.

Proof. Since the proofs of (a) and (b) are similar, we only prove (a). Since A is strongly sublinear, for any $x > \theta$, we have

$$Ax = A(\frac{1}{2} \cdot 2x) >> \frac{1}{2}A(2x) \geq 0,$$

and hence $Ax \in \overset{o}{P}$. As a result, there exists sufficiently small $\alpha > 0$ such that $Ax - \alpha e \in P$, i.e., $Ax \geq \alpha e$. In the same way, by $e \in \overset{o}{P}$ we can choose sufficiently large $\beta > 0$ such that $Ax \leq \beta e$. Hence (2.2.1) is satisfied. Now, for any $x > \theta$ and $0 < t < 1$, since $A(tx) - tAx \in \overset{o}{P}$, we can choose sufficiently small $\eta > 0$ such that $A(tx) - tAx - \eta tAx \in P$, i.e., $A(tx) \geq (1 + \eta)tAx$, and hence (2.2.2) is satisfied, which implies A is e-concave. □

Conclusions (a) and (b) of Theorem 2.2.1 are not true for $e \in \partial P$, $e \neq \theta$. For example, the operator (2.2.8) is e-concave, but it is not strongly sublinear, since for any $u > \theta$ and $0 < t < 1$, we have $A(tu(x)) - tAu(x) = 0$, $\forall\ x \in \partial\Omega$ and hence $A(tu) - tAu \notin \overset{o}{P}$.

We shall se that e-concave and sublinear operators have only weak nonlinearity, making it easy to deal with them. However, e-convex and superlinear operators always have strong nonlinearity, and there are only a few results about them.

Theorem 2.2.2. If operator $A : P \to P$ is increasing and e-concave, then A has at most one positive (i.e., $> \theta$) fixed point.

Proof. Suppose that $x_1 > \theta$ and $x_2 > \theta$ are two positive fixed points of A. Then, by (2.2.1), we see

$$x_1 = Ax_1 \geq \alpha_1 e = \frac{\alpha_1}{\beta_2} \cdot \beta_2 e \geq \frac{\alpha_1}{\beta_2} Ax_2 = \frac{\alpha_1}{\beta_2} x_2 ,$$

where α_1 and β_2 are positive constants. Letting $t_0 = \sup\{t > 0 \mid x_1 \geq tx_2\}$, we see that $0 < t_0 < +\infty$. Now we prove $t_0 \geq 1$. In fact, if $0 < t_0 < 1$, then, by (2.2.2) there exists $\eta_0 > 0$ such that

$$x_1 = Ax_1 \geq A(t_0x_2) \geq (1 + \eta_0)t_0Ax_2 = (1 + \eta_0)t_0x_2,$$

which contradicts the definition of t_0. Hence $t_0 \geq 1$ and so $x_1 \geq x_2$. In the same way, we can prove $x_2 \geq x_1$. Thus, $x_2 = x_1$. □

Theorem 2.2.3. Let operator $A : P \to P$ be increasing and e-concave. Suppose that A has positive fixed point $x^* > \theta$ and cone P is normal. Then there exist $R > r > 0$ such that

$$Ax \nleq x, \quad \forall\, x \in P, \quad 0 < \|x\| < r, \tag{2.2.16}$$

$$Ax \ngeq x, \quad \forall\, x \in P, \quad \|x\| > R. \tag{2.2.17}$$

Proof. We first prove

$$x > \theta, \quad Ax \leq x \Rightarrow x \geq x^* . \tag{2.2.18}$$

In fact, from (2.2.1) we know

$$x \geq Ax \geq \alpha e = \frac{\alpha}{\beta^*} \cdot \beta^* e \geq \frac{\alpha}{\beta^*} Ax^* = \frac{\alpha}{\beta^*} x^*,$$

where α, β^* are positive numbers. Hence, putting $t_0 = \sup\{t > 0 \mid x \geq tx^*\}$, we see $0 < t_0 < +\infty$. We now prove $t_0 \geq 1$. If $0 < t_0 < 1$, then we know by (2.2.2) that there exists $\eta_0 > 0$ such that

$$x \geq Ax \geq A(t_0x^*) \geq (1 + \eta_0)t_0Ax^* = (1 + \eta_0)t_0x^*,$$

in contradiction to the definition of t_0. Hence $t_0 \geq 1$ and (2.2.18) is proved.

Similarly, we can prove

$$x > \theta, \quad Ax \geq x \Rightarrow x \leq x^* . \tag{2.2.19}$$

Now, let $r = \inf_{z \in P} \|z + x^*\|$. Since $x^* > \theta$, it is easy to see that $r > 0$. We prove this r satisfies (2.2.16). In fact, if $Ax \leq x$ for some $x \in P$, $0 \leq \|x\| < r$, then by (2.2.18) $x = z + x^*$, where $z = x - x^* \geq \theta$, and therefore $\|x\| \geq r$, a contradiction.

Since P is normal, $[\theta, x^*]$ is bounded. We now prove that the number $R = \sup_{\theta \leq z \leq x^*} \|z\|$ satisfies (2.2.17). In fact, if $Ax \geq x$ for some $x \in P$, $\|x\| > R$, then by (2.2.19) we have $x \in [\theta, x^*]$, in contradiction to the definition of R. □

When A satisfies inequalities (2.2.16) and (2.2.17), we say that A is <u>cone compression</u>. In the next section we shall see that if cone compression mapping A is completely continuous, then A has a positive fixed point. Thus, an increasing and completely continuous e-concave operator has positive fixed point if and only if A is cone compression.

<u>Theorem 2.2.4</u>. Let operator $A : P \to P$ be increasing and e-concave. Suppose that A has positive fixed point $x^* > 0$. Then, constructing successively the sequence $x_n = Ax_{n-1}$ $(n = 1,2,3,\ldots)$ for any initial $x_0 > \theta$, we have

$$\|x_n - x^*\|_e \to 0 \quad (n \to \infty). \tag{2.2.20}$$

<u>Proof</u>. For any $0 < t_1 < 1$, let

$$v_0 = t_1 x^*, \quad v_{n+1} = Av_n \quad (n = 0,1,2,\ldots).$$

Evidently,

$$t_1 x^* = v_0 \leq v_1 \leq \ldots \leq v_n \leq \ldots \leq x^* . \tag{2.2.21}$$

Let $\rho_n = \sup\{t > 0 \mid tx^* \leq v_n\}$, then

$$0 < t_1 = \rho_0 \leq \rho_1 \leq \rho_2 \leq \cdots \leq \rho_n \leq \cdots \leq 1, \tag{2.2.22}$$

and

$$\rho_n x^* \leq v_n \quad (n = 0,1,2,\ldots). \tag{2.2.23}$$

We now prove

$$\lim_{n\to\infty} \rho_n = 1. \tag{2.2.24}$$

If otherwise, $\lim_{n\to\infty} \rho_n = \gamma < 1$ $(\gamma \geq t_1 > 0)$. Then there exists $\eta > 0$ such that

$$A(\gamma x^*) \geq (1 + \eta)\gamma A x^* = (1 + \eta)\gamma x^*,$$

hence, for $0 < t \leq \gamma$,

$$A(tx^*) = A(\frac{t}{\gamma} \cdot \gamma x^*) \geq \frac{t}{\gamma} A(\gamma x^*) \geq (1 + \eta) t x^*,$$

in particular,

$$A(\rho_n x^*) \geq (1 + \eta)\rho_n x^* \quad (n = 0,1,2,\ldots). \tag{2.2.25}$$

It follows from (2.2.23) and (2.2.25) that

$$v_{n+1} = Av_n \geq A(\rho_n x^*) \geq (1 + \eta)\rho_n x^* ,$$

and therefore,

$$\rho_{n+1} \geq (1 + \eta)\rho_n \quad (n = 0,1,2,\ldots),$$

which in turn yields,

$$\rho_n \geq (1 + \eta)^n \rho_0 \quad (n = 0,1,2,\ldots),$$

in contradiction with (2.2.22). Thus, (2.2.24) holds.

Now, for any $t_2 > 1$, let

$$w_0 = t_2 x^*, \quad w_{n+1} = Aw_n \quad (n = 0,1,2,\ldots).$$

It is easy to see that

$$t_2 x^* = w_0 \geq w_1 \geq \cdots \geq w_n \geq \cdots \geq x^*, \tag{2.2.26}$$

$$t_2 \geq \xi_0 \geq \xi_1 \geq \cdots \geq \xi_n \geq \cdots \geq 1, \tag{2.2.27}$$

and

$$\xi_n x^* \geq w_n \quad (n = 0,1,2,\ldots), \tag{2.2.28}$$

where $\xi_n = \inf\{t > 0 \mid tx^* \geq w_n\}$ $(n = 0,1,2,\ldots)$. We now prove

$$\lim_{n\to\infty} \xi_n = 1. \tag{2.2.29}$$

If otherwise, $\lim\limits_{n\to\infty} \xi_n = \gamma_0 > 1$. Then there exists $\eta_0 > 0$ such that

$$x^* = Ax^* = A(\frac{1}{\gamma_0} \cdot \gamma_0 x^*) \geq \frac{1}{\gamma_0} \ (1 + \eta_0)A(\gamma_0 x^*),$$

i.e.,

$$A(\gamma_0 x^*) \leq \frac{\gamma_0}{1 + \eta_0} \ x^*,$$

and hence, for $t \geq \gamma_0$ we have

$$A(\gamma_0 x^*) = A(\frac{\gamma_0}{t} \cdot tx^*) \geq \frac{\gamma_0}{t} A(tx^*) \ ,$$

and so

$$A(tx^*) \leq \frac{t}{\gamma_0} A(\gamma_0 x^*) \leq \frac{t}{1 + \eta_0} x^* \ ,$$

in particular,

$$A(\xi_n x^*) \leq \frac{\xi_n x^*}{1 + \eta_0} \quad (n = 0,1,2,\ldots). \tag{2.2.30}$$

It follows from (2.2.28) and (2.2.30) that

$$w_{n+1} \leq A(\xi_n x^*) \leq \frac{\xi_n}{1 + \eta_0} x^* \ ,$$

and hence

$$\xi_{n+1} \le \frac{\xi_n}{1 + \eta_0} \quad (n = 0,1,2,\ldots).$$

Consequently,

$$\xi_n \le \frac{\xi_0}{(1 + \eta_0)^n} \quad (n = 0,1,2,\ldots),$$

which contradicts (2.2.27), and so (2.2.29) holds.

Now, let $x_0 > \theta$ be given and construct successively the sequence $x_n = Ax_{n-1}$ $(n = 1,2,3,\ldots)$. By (2.2.1), we have

$$\begin{aligned} \alpha_0 e &\le x^* = Ax^* \le \beta_0 e\ , \\ \alpha_1 e &\le x_1 = Ax_0 \le \beta_1 e\ , \end{aligned} \tag{2.2.31}$$

where $\alpha_0, \beta_0, \alpha_1$ and β_1 are positive numbers. Hence,

$$\alpha_1 \beta_0^{-1} x^* \le x_1 \le \beta_1 \alpha_0^{-1} x^*\ .$$

Now, we choose t_1 and t_2 such that

$$0 < t_1 < \min\{1, \alpha_1 \beta_0^{-1}\}, \quad t_2 > \max\{1, \beta_1 \alpha_0^{-1}\}.$$

Then

$$0 < t_1 < 1, \quad t_2 > 1, \quad v_0 = t_1 x^* \le x_1 \le t_2 x^* = w_0,$$

and by (2.2.23) and (2.2.28),

$$\rho_n x^* \le v_n \le x_{n+1} \le w_n \le \xi_n x^* \quad (n = 0,1,2,\ldots).$$

Hence, noticing (2.2.31),

$$-(1 - \rho_n)\beta_0 e \le x_{n+1} - x^* \le (\xi_n - 1)\beta_0 e \quad (n = 0,1,2,\ldots). \tag{2.2.32}$$

Finally, (2.2.20) follows from (2.2.32), (2.2.24), and (2.2.29). □

Corollary 2.2.1. If P is normal and the hypotheses of Theorem 2.2.4 are satisfied, then $\|x_n - x^*\| \to 0 \quad (n \to \infty)$.

The next theorem is concerned with e-convex operators. In this theorem we shall use the so-called e-increasing operator. Let $e > \theta$. Operator $A : P \to P$ is said to be e-increasing if $\theta \leq x < y$ implies $Ay - Ax \geq \alpha e$ for some $\alpha = \alpha(x,y) > 0$. Evidently, (a) if A is strongly increasing in case $\overset{o}{P} \neq \phi$, then A is e-increasing for any $e > \theta$ and (b) if A is e-increasing for some $e > \theta$, then A is strictly increasing. Similarly, we can define the e-decreasing operators and obtain corresponding results.

Theorem 2.2.5. If operator $A : P \to P$ is e-convex and e-increasing then A cannot have two comparable positive fixed points.

Proof. Suppose that A has two positive fixed points $x_1 > \theta$ and $x_2 > \theta$, which are comparable, for example, $x_1 < x_2$. Since A is e-increasing, there exists an $\alpha > 0$ such that

$$x_2 - x_1 = Ax_2 - Ax_1 \geq \alpha e .$$

By (2.2.1), we have

$$\alpha_2 e \leq x_2 = Ax_2 \leq \beta_2 e ,$$

where $\alpha_2 > 0$, $\beta_2 > 0$, and hence

$$x_1 \leq (1 - \frac{\alpha}{\beta_2})x_2 .$$

Now, let $t_0 = \inf\{t > 0 \mid x_1 \leq tx_2\}$. From (2.2.33) we see that

$$0 < t_0 \leq 1 - \frac{\alpha}{\beta_2} < 1, \quad x_1 \leq t_0 x_2 .$$

Thus, by e-convexity of A, there exists an $\eta > 0$ such that

$$x_1 = Ax_1 \leq A(t_0x_2) \leq (1 - \eta)t_0Ax_2 = (1 - \eta)t_0x_2,$$

which contradicts the definition of t_0. □

Notice, under the hypotheses of Theorem 2.2.5, A may have many positive fixed points. Of course they are not comparable to each other. For example, let $E = R^2$, $P = \{x = (x_1,x_2) \in R^2 \mid x_1 \geq 0,\ x_2 \geq 0\}$, and

$$Ax = (x_1^2 + x_2,\ \frac{1}{16}x_1 + 4x_2^2),\quad x = (x_1,x_2).$$

It is easy to verify that operator $A : P \to P$ is e-convex and e-increasing where $e = (1,1)$, but A has three positive fixed points $x^{(1)}, x^{(2)}$, and $x^{(3)}$:

$$x^{(1)} = \left(\frac{3}{4},\frac{3}{16}\right),\quad x^{(2)} = \left(\frac{5 + \sqrt{5}}{8},\frac{5 - \sqrt{5}}{32}\right),\quad x^{(3)} = \left(\frac{5 - \sqrt{5}}{8},\frac{5 + \sqrt{5}}{32}\right).$$

Of course, $x^{(1)}, x^{(2)}$, and $x^{(3)}$ are not comparable.

Under what conditions an e-convex operator (in particular, strongly superlinear operator) has exactly one positive fixed point is an open problem.

Example 2.2.3. We consider the nonlinear integral operator (2.2.6) of Example 2.2.1 again.

(a) Let $0 < \alpha < 1$. In this case, we already know from Example 2.2.1 that A is strongly sublinear and therefore, by Theorem 2.2.1, A is e-concave for any $e \in \overset{o}{P}$. On the other hand, it is easy to verify that A is strongly increasing for any $\alpha > 0$ (not only for $0 < \alpha < 1$). Hence, by Theorem 2.2.2, A has at most one positive fixed point.

Now, let

$$u_0(x) \equiv (m \text{ mes } G)^{\frac{1}{1-\alpha}}, \quad v_0(x) \equiv (M \text{ mes } G)^{\frac{1}{1-\alpha}}, \quad \forall\, x \in G,$$

where

$$m = \min_{x,y \in G} k(x,y) > 0, \quad M = \max_{x,y \in G} k(x,y) > 0.$$

We have

$$Au_0(x) = \int_G k(x,y)[u_0(y)]^\alpha dy \geq (m \text{ mes } G)^{\frac{1}{1-\alpha}} = u_0(x),$$

$$Av_0(x) = (M \text{ mes } G)^{\frac{\alpha}{1-\alpha}} \int_G k(x,y)dy \leq (M \text{ mes } G)^{\frac{1}{1-\alpha}} = v_0(x),$$

and hence, by Theorem 2.1.1, A has at least one fixed point ψ^* such that $u_0 \leq \psi^* \leq v_0$.

Thus, A has exactly one positive fixed point ψ^* and $\psi^*(x) > 0$, $\forall\, x \in G$. Moreover, by Theorem 2.2.4 and Corollary 2.2.1, for any initial function $\psi_0(x)$ which is nonnegative and not identically zero on G, constructing successively functions

$$\psi_n(x) = A\psi_{n-1}(x) = \int_G k(x,y)[\psi_{n-1}(y)]^\alpha dy \quad (n = 1,2,3,\ldots),$$

the sequence $\{\psi_n(x)\}$ must converge to $\psi^*(x)$ uniformly on G.

(b) Let $\alpha > 1$. We already know that A is strongly superlinear and strongly increasing, and therefore A is e-convex and e-increasing for any $e \in \overset{o}{P}$. Hence, by Theorem 2.2.5, A cannot have two comparable positive fixed points.

In the next section, using the fixed point theorem of cone expansion, we shall show that A has at least one positive fixed point. But, we cannot assert that A has exactly one positive fixed point.

<u>Definition 2.2.3</u>. Let P be solid and $A : \overset{o}{P} \to \overset{o}{P}$. Let $0 \leq \alpha < 1$. A is said to be <u>α-concave</u> if

$$A(tx) \geq t^\alpha Ax, \ \forall\ x \in \overset{o}{P}, \quad 0 < t < 1. \tag{2.2.34}$$

Similarly, A is said to be <u>$(-\alpha)$-convex</u> if

$$A(tx) \leq t^{-\alpha}Ax, \quad \forall\, x \in \overset{o}{P}, \quad 0 < t < 1. \tag{2.2.35}$$

It is easy to show that A is α-concave if and only if

$$A(sx) \leq s^{\alpha}Ax, \quad \forall\, x \in \overset{o}{P}, \quad s > 1. \tag{2.2.36}$$

Similarly, A is $(-\alpha)$-convex if and only if

$$A(sx) \geq s^{-\alpha}Ax, \quad \forall\, x \in \overset{o}{P}, \quad s > 1. \tag{2.2.37}$$

<u>Theorem 2.2.6</u>. Let P be normal and operator $A : \overset{o}{P} \to \overset{o}{P}$ be α-concave and increasing or $(-\alpha)$-convex and decreasing $(0 \leq \alpha < 1)$. Then, A has exactly one fixed point x^* in $\overset{o}{P}$. Moreover, constructing successively the sequence $x_n = Ax_{n-1}$ $(n = 1,2,3,\ldots)$ for any initial $x_0 \in \overset{o}{P}$, we have

$$\|x_n - x^*\| \to 0 \quad (n \to \infty) \tag{2.2.38}$$

and the rate of convergence is

$$\|x_n - x^*\| = 0(1 - \gamma^{\alpha^n}), \tag{2.2.39}$$

where $0 < \gamma < 1$, which depends on initial element x_0.

<u>Proof</u>. First we assume that A is α-concave and increasing. Let $x_0 \in \overset{o}{P}$ be given. Choose t_0 and s_0 such that

$$0 < t_0 < 1 < s_0, \quad t_0^{1-\alpha}x_0 \leq Ax_0 \leq s_0^{1-\alpha}x_0,$$

and let $u_0 = t_0x_0$, $v_0 = s_0x_0$, then $u_0 \ll v_0$ and

$$Au_0 \geq t_0^{\alpha}Ax_0 \geq t_0x_0 = u_0, \quad Av_0 \leq s_0^{\alpha}\, Ax_0 \leq s_0x_0 = v_0. \tag{2.2.40}$$

Now we define

$$u_n = Au_{n-1}, \quad v_n = Av_{n-1} \quad (n = 1,2,3,\ldots).$$

It is easy to prove from (2.2.40) that

$$u_0 \leq u_1 \leq \cdots \leq u_n \leq \cdots \leq v_n \leq \cdots \leq v_1 \leq v_0. \tag{2.2.41}$$

Clearly, $u_0 = \gamma v_0$, where $\gamma = t_0/s_0$, $0 < \gamma < 1$. By induction, we see easily that

$$u_n \geq \gamma^{\alpha^n} v_n \, (n = 0,1,2,\ldots). \tag{2.2.42}$$

It follows from (2.2.41) and (2.2.42) that

$$\theta \leq u_{n+p} - u_n \leq v_n - u_n \leq (1 - \gamma^{\alpha^n})v_n \leq (1 - \gamma^{\alpha^n})v_0, \tag{2.2.43}$$

and therefore, by the normality of P and Theorem 1.1.1, u_n converges to some $u^* \in E$. In the same way, we can prove that $v_n \to v^* \in E$. By (2.2.41), we see

$$u_n \leq u^* \leq v^* \leq v_n \quad (n = 0,1,2,\ldots), \tag{2.2.44}$$

and so

$$u_{n+1} = Au_n \leq Au^* \leq Av^* \leq Av_n = v_{n+1} .$$

Taking limit $n \to \infty$, we get

$$u^* \leq Au^* \leq Av^* \leq v^*. \tag{2.2.45}$$

Now, from (2.2.43) and (2.2.44), we obtain

$$\theta \leq v^* - u^* \leq v_n - u_n \leq (1 - \gamma^{\alpha^n})v_0,$$

which implies $v^* = u^*$, and therefore by (2.2.45)

$$u^* = Au^* = Av^* = v^*. \tag{2.2.46}$$

Now, setting $x^* = u^* = v^*$, we see that x^* is a fixed point of A in $\overset{o}{P}$. It follows from $u_0 \le x_0 \le v_0$ that $u_n \le x_n \le v_n$ $(n = 1,2,3,\ldots)$, and so by (2.2.43) and (2.2.44)

$$\|x_n - x^*\| \le \|x_n - u_n\| + \|u_n - x^*\| \le 2N\|v_n - u_n\|$$
$$\le 2N^2(1 - \gamma^{\alpha^n})\|v_0\|,$$

where N is the normal constant of P. Hence, (2.2.39) and (2.2.38) hold. It remains to prove the uniqueness of fixed points of A in $\overset{o}{P}$. Let $\bar{x} \in \overset{o}{P}$ be any fixed point of A. Put $t_1 = \sup\{t > 0 \mid \bar{x} \ge tx^*\}$. Evidently $0 < t_1 < +\infty$. We now prove $t_1 \ge 1$. In fact, if $0 < t_1 < 1$, then

$$\bar{x} = A\bar{x} \ge A(t_1 x^*) \ge t_1^{\alpha} Ax^* = t_1^{\alpha} x^* ,$$

which contradicts the definition of t_1 since $t_1^{\alpha} > t_1$. Thus $t_1 \ge 1$ and $\bar{x} \ge x^*$. In the same way, we can prove $x^* \ge \bar{x}$ and hence $\bar{x} = x^*$, the uniqueness of fixed points of A in $\overset{o}{P}$ is proved.

Next, we assume that A is $(-\alpha)$-convex and decreasing. It is easy to see that operator $A^2 : \overset{o}{P} \to \overset{o}{P}$ is α^2-concave and increasing. Hence, by the result proved above, A^2 has exactly one fixed point x^* in $\overset{o}{P}$ and $z_n = A^{2n} z_0 \to x^*$ for any initial $z_0 \in \overset{o}{P}$ with rate of convergence:

$$\|z_n - x^*\| = O(1 - \gamma^{\alpha^{2n}}), \tag{2.2.47}$$

where $0 < \gamma < 1$. Since $A^2(Ax^*) = A(A^2 x^*) = Ax^*$, by uniqueness, we have $Ax^* = x^*$. Choosing $z_0 = x_0$ and $z_0 = Ax_0$ in (2.2.47), respectively, we obtain $x_{2n} = A^{2n} x_0 \to x^*$ and $x_{2n+1} = A^{2n+1} x_0 \to x^*$ and

$$\|x_{2n} - x^*\| = O(1 - \gamma_1^{\alpha^{2n}}), \quad 0 < \gamma_1 < 1,$$

$$\|x_{2n+1} - x^*\| = O(1 - \gamma_2^{\alpha^{2n}}) = O(1 - (\gamma_2^{1/\alpha})^{\alpha^{2n+1}}), \quad 0 < \gamma_2 < 1.$$

Hence (2.2.39) holds, where $\gamma = \min\{\gamma_1, \gamma_2^{1/\alpha}\}$. □

Notice that we do not assume A to be continuous in Theorem 2.2.6.

If, in Theorem 2.2.6, we replace the assumptions "α-concave" and "$(-\alpha)$-convex" by their particular cases "α-homogeneous" (i.e., $A(tx) = t^{\alpha}Ax$, $\forall\, x \in \overset{o}{P}$, $0 < t < 1$) and "$(-\alpha)$-homogeneous" (i.e., $A(tx) = t^{-\alpha}Ax$, $\forall\, x \in \overset{o}{P}$, $0 < t < 1$), respectively, then we can use Hilbert's projective metric to prove the existence and uniqueness of fixed points of A in $\overset{o}{P}$, as follows.

First assume that A is α-homogeneous and increasing. Since $t > 1$ implies $Ax = A(t^{-1} \cdot tx) = t^{-\alpha}A(tx)$, we see $A(tx) = t^{\alpha}Ax$ for any $x \in \overset{o}{P}$ and $t > 0$. For any $x, y \in \overset{o}{P}$, we have

$$m\left(\frac{x}{y}\right)y \le x \le M\left(\frac{x}{y}\right)y ,$$

and hence

$$\left[m\left(\frac{x}{y}\right)\right]^{\alpha} Ay = A\left[m\left(\frac{x}{y}\right)y\right] \le Ax \le A\left[M\left(\frac{x}{y}\right)y\right] = \left[M\left(\frac{x}{y}\right)\right]^{\alpha} Ay .$$

Therefore

$$M\left(\frac{Ax}{Ay}\right) \le \left[M\left(\frac{x}{y}\right)\right]^{\alpha}, \quad m\left(\frac{Ax}{Ay}\right) \ge \left[m\left(\frac{x}{y}\right)\right]^{\alpha},$$

and thus

$$d(Ax,Ay) = \ln \frac{M\left(\frac{Ax}{Ay}\right)}{m\left(\frac{Ax}{Ay}\right)} \le \alpha \ln \frac{M\left(\frac{x}{y}\right)}{m\left(\frac{x}{y}\right)} = \alpha\, d(x,y). \tag{2.2.48}$$

Now define

$$A_1 x = \frac{Ax}{\|Ax\|} , \tag{2.2.49}$$

then $A_1 : \overset{o}{P} \cap S_1 \to \overset{o}{P} \cap S_1$, where $S_1 = \{x \in E \mid \|x\| = 1\}$, and for any $x, y \in \overset{o}{P} \cap S_1$, we have

$$d(A_1x, A_1y) = d(\|Ax\|^{-1}Ax, \|Ay\|^{-1}Ay) = d(Ax, Ay) \leq \alpha\, d(x,y).$$

Consequently, by Theorem 1.5.2 and Banach's fixed point theorem of contraction mapping, A_1 has exactly one fixed point $x_1 \in \overset{o}{P} \cap S_1$. Let $x^* = \|Ax_1\|^{1/1-\alpha}x_1$; we have

$$Ax^* = \|Ax_1\|^{\alpha/1-\alpha}Ax_1 = \|Ax_1\|^{\alpha/1-\alpha+1}A_1x_1 = x^*.$$

On the other hand, if $y^* \in \overset{o}{P}$ such that $Ay^* = y^*$, then from (2.2.48) we find

$$d(x^*,y^*) = d(Ax^*,Ay^*) \leq \alpha\, d(x^*,y^*).$$

As a result, $d(x^*,y^*) = 0$, and so $x^* = \lambda y^*$ for some $\lambda > 0$. Thus,

$$x^* = Ax^* = A(\lambda y^*) = \lambda^{\alpha}Ay^* = \lambda^{\alpha}y^* ,$$

and therefore $\lambda = 1$ and $x^* = y^*$. This proves that A has exactly one fixed point x^* in $\overset{o}{P}$.

Next we assume that A is $(-\alpha)$-homogeneous and decreasing. In this case, $m(x/y)y \leq x \leq M(x/y)y$ implies

$$\left[M\left(\frac{x}{y}\right)\right]^{-\alpha} Ay \leq Ax \leq \left[m\left(\frac{x}{y}\right)\right]^{-\alpha} Ay ,$$

and therefore

$$d(Ax,Ay) \leq \ln \frac{\left[m\left(\frac{x}{y}\right)\right]^{-\alpha}}{\left[M\left(\frac{x}{y}\right)\right]^{-\alpha}} = \alpha\, d(x,y) .$$

Similarly, considering the operator defined in (2.2.49), we can prove that A_1 has exactly one fixed point x_1 in $\overset{o}{P} \cap S_1$ and $x^* = \|Ax_1\|^{1/1+\alpha}x_1$ is the unique fixed point of A in $\overset{o}{P}$.

Theorem 2.2.7. Let P be normal and operator $A : \mathring{P} \to \mathring{P}$ be α-concave and increasing or $(-\alpha)$-convex and decreasing. Denote the unique solution in $\mathring{P}$ of the equation $Ax = \lambda x$ by x_λ. Then x_λ is strongly decreasing (i.e., $0 < \lambda_1 < \lambda_2$ implies $x_{\lambda_1} \gg x_{\lambda_2}$), continuous (i.e., $\lambda \to \lambda_0$, $\lambda_0 > 0$ implies $\|x_\lambda - x_{\lambda_0}\| \to 0$), and

$$\lim_{\lambda\to+\infty} \|x_\lambda\| = 0, \quad \lim_{\lambda\to+0} \|x_\lambda\| = +\infty . \tag{2.2.50}$$

Proof. We need only to prove the theorem for the case that A is α-concave and increasing, given the similarity of the two.

By Theorem 2.2.6, equation $Ax = \lambda x$ $(\lambda > 0)$ has exactly one solution $x_\lambda \in \mathring{P}$. Let $0 < \lambda_1 < \lambda_2$ and $t_0 = \sup\{t > 0 \mid x_{\lambda_1} \geq t x_{\lambda_2}\}$. Then $0 < t_0 < +\infty$. We now prove $t_0 \geq 1$. In fact, if $0 < t_0 < 1$, then

$$x_{\lambda_1} = \frac{1}{\lambda_1} Ax_{\lambda_1} \geq \frac{1}{\lambda_1} A(t_0 x_{\lambda_2}) \geq \frac{t_0^\alpha}{\lambda_1} Ax_{\lambda_2} = \lambda_1^{-1}\lambda_2 t_0^\alpha x_{\lambda_2} ,$$

which contradicts the definition of t_0 since $\lambda_1^{-1}\lambda_2 t_0^\alpha > t_0$. Hence $t_0 \geq 1$ and $x_{\lambda_1} \geq x_{\lambda_2}$, and therefore

$$x_{\lambda_1} = \frac{1}{\lambda_1} Ax_{\lambda_1} \geq \frac{1}{\lambda_1} Ax_{\lambda_2} = \frac{\lambda_2}{\lambda_1} x_{\lambda_2} \gg x_{\lambda_2}, \quad \forall\ 0 < \lambda_1 < \lambda_2. \tag{2.2.51}$$

Thus A is strongly decreasing. Letting $\lambda_1 = 1$ and $\lambda_2 = \lambda$ in (2.2.51), we see $x_1 \geq \lambda x_\lambda$, $\forall\ \lambda > 1$, and so

$$\|x_\lambda\| \leq \frac{N}{\lambda} \|x_1\|, \quad \forall\ \lambda > 1, \tag{2.2.52}$$

where N denotes the normal constant of P. Hence $\|x_\lambda\| \to 0$ as $\lambda \to +\infty$. Similarly, letting $\lambda_1 = \lambda$ and $\lambda_2 = 1$ in (2.2.51), we can get $\|x_\lambda\| \to +\infty$ as $\lambda \to +0$. Thus, (2.2.50) holds. Finally, we prove the continuity of x_λ. Let $\lambda_0 > 0$ be given. By (2.2.51) we know

$$x_\lambda \ll x_{\lambda_0} \quad \forall\ \lambda > \lambda_0 . \tag{2.2.53}$$

On the other hand, letting $\gamma_\lambda = \sup\{t > 0 \mid x_\lambda \geq t x_{\lambda_0}\}$, $(\lambda > \lambda_0)$, we see $0 < \gamma_\lambda < 1$, and hence

$$\lambda x_\lambda = A x_\lambda \geq A(\gamma_\lambda x_{\lambda_0}) \geq \gamma_\lambda^\alpha A x_{\lambda_0} = \gamma_\lambda^\alpha \lambda_0 x_{\lambda_0} .$$

Therefore, by the definition of γ_λ, $\lambda^{-1}\gamma_\lambda^\alpha \lambda_0 \leq \gamma_\lambda$, i.e.,

$$\gamma_\lambda \geq \left(\frac{\lambda_0}{\lambda}\right)^{1/1-\alpha}$$

Thus

$$x_\lambda \geq \left(\frac{\lambda_0}{\lambda}\right)^{1/1-\alpha} x_{\lambda_0}, \quad \forall \lambda > \lambda_0 . \tag{2.2.54}$$

It follows from (2.2.53) and (2.2.54) that

$$\|x_{\lambda_0} - x_\lambda\| \leq N\left[1 - \left(\frac{\lambda_0}{\lambda}\right)^{1/1-\alpha}\right]\cdot\|x_{\lambda_0}\| \to 0 \quad \text{as} \quad \lambda \to \lambda_0 + 0 .$$

In the same way, we can prove $\|x_\lambda - x_{\lambda_0}\| \to 0$ as $\lambda \to \lambda_0 - 0$. Hence, x_λ is continuous at $\lambda = \lambda_0$.

Notice that even though we do not assume A is continuous, we can assert that x_λ depends continuously on λ.

In the following, for any $x_0 \in \overset{o}{P}$ with $0 < \|x_0\| < 1$, we define

$$P(x_0) = \{x \in P \mid x \geq \|x\| x_0\}.$$

Obviously, $P(x_0)$ is a cone of E and $P(x_0)\backslash\{\theta\} \subset \overset{o}{P}$. Moreover, for any $x \in \overset{o}{P}$ there exists $x_0 \in \overset{o}{P}$ with $0 < \|x_0\| < 1$ such that $x \in P(x_0)$.

Theorem 2.2.8. Let P be normal. The following conclusions hold:

(a) If $A : \overset{o}{P} \to \overset{o}{P}$ is α-concave and increasing, then A is cone compression with respect to $\overset{o}{P}$, i.e., there exists $R > r > 0$ such that

$$Ax \not\leq x, \quad \forall x \in \overset{o}{P}, \quad 0 < \|x\| < r, \tag{2.2.55}$$

$$Ax \not\geq x, \quad \forall x \in \overset{o}{P}, \quad \|x\| > R. \tag{2.2.56}$$

(b) If $A : \overset{o}{P} \to \overset{o}{P}$ is $(-\alpha)$-convex and decreasing, then for any $x_0 \in \overset{o}{P}$ with $0 < \|x_0\| < 1$, A is strong cone compression with respect to $P(x_0)$, i.e., there exist $R > r > 0$ (R depends on x_0 and r is independent of x_0) such that

$$Ax \gg x, \ \forall \ x \in \overset{o}{P}, \quad 0 < \|x\| < r, \tag{2.2.57}$$

$$Ax \ll x, \ \forall \ x \in P(x_0), \quad \|x\| > R. \tag{2.2.58}$$

Proof. The proof of (a) is similar to the proof of Theorem 2.2.3. We only need to replace (2.2.18) and (2.2.19) by

$$x \in \overset{o}{P}, \quad Ax \leq x \Rightarrow x \geq x^*$$

and

$$x \in \overset{o}{P}, \quad Ax \geq x \Rightarrow x \leq x^* ,$$

respectively, where $x^* \in \overset{o}{P}$ denotes the unique fixed point of A.

Now, we prove conclusion (b). By Theorem 2.2.6, A has exactly one fixed point $x^* \in \overset{o}{P}$. We can choose $r > 0$ such that

$$x^* - x \in \overset{o}{P}, \ \forall \ x \in \overset{o}{P}, \quad 0 < \|x\| < r. \tag{2.2.59}$$

Then, this r satisfies (2.2.57). In fact, $x \in \overset{o}{P}$ with $0 < \|x\| < r$ implies $x \ll x^*$, and therefore $Ax \geq Ax^* = x^* \gg x$.

Let $x_0 \in \overset{o}{P}$ with $0 < \|x_0\| < 1$ be given. Choosing $R > r$ such that $x_0 - x^*/R \in \overset{o}{P}$, we see that this R satisfies (2.2.58). In fact, $x \in P(x_0)$ with $\|x\| > R$ implies

$$x - x^* = \|x\| \cdot \left(\frac{x}{\|x\|} - \frac{x^*}{\|x\|}\right) \geq \|x\| \cdot \left(x_0 - \frac{x^*}{R}\right) \gg \theta ,$$

and hence

$$Ax \leq Ax^* = x^* \ll x \ . \ \square$$

We cannot strengthen (2.2.58) to the following form:

$$Ax \ll x, \quad \forall\, x \in \mathring{P}, \quad \|x\| > R \ . \tag{2.2.60}$$

For example, consider the integral operator

$$Ax(t) = \int_0^1 k(t,x)\; x(s)^{-\alpha} ds \ , \tag{2.2.61}$$

where $0 < \alpha < 1$, $k(t,s)$ is continuous and positive on $0 \leq t, \ s \leq 1$. Let $E = C\ 0,1$, $P = \{x(t) \in C[0,1] \mid x(t) \geq 0\}$. Obviously, $A : \mathring{P} \to \mathring{P}$ is $(-\alpha)$-convex and decreasing. Define

$$x_n(t) = \begin{cases} -(n^2 - 1)t + n, & 0 \leq t \leq \frac{1}{n} \ ; \\ \frac{1}{n} \ , & \frac{1}{n} < t \leq 1 \ . \end{cases}$$

It is evident that $x_n(t) \in \mathring{P}$ and $\|x_n\| = n$ $(n = 2,3,4,\ldots)$. Denoting

$$\tau = \min_{0 \leq t,s \leq 1} k(t,s) > 0 \ ,$$

we have

$$Ax_n(1) = \int_0^1 k(1,s)\; x_n(s)^{-\alpha} ds \geq \int_{\frac{1}{2}}^1 \tau \left(\frac{1}{n}\right)^{-\alpha} ds$$

$$= \frac{\tau}{2}\, n^{\alpha} = \frac{\tau}{2}\, n^{\alpha+1} x_n(1) \ .$$

Hence $Ax_n(1) > x_n(1)$ for sufficiently large n, and therefore (2.2.60) does not hold for any $R > 0$.

<u>Example 2.2.4</u>. Consider in Euclidean space R^n two Hammerstein integral equations:

$$\lambda x(t) = \int_{R^n} k(t,s) \cdot \left\{ \sum_{i=1}^{\infty} a_i(s)\; x(s)^{\alpha_i} \right\} ds \tag{2.2.62}$$

and

$$x(t) = \int_{R^n} k(t,s)\cdot\left\{ \sum_{i=1}^{\infty} a_i(s)\, x(s)^{\alpha_i}\right\}^{-1} ds \,. \tag{2.2.63}$$

Suppose that (i) $\alpha_i > 0$ $(i = 1,2,3,\ldots)$ and $\sup_i \alpha_i < 1$; (ii) kernel $k(t,s)$ is measurable and nonnegative on R^{2n} and there exist constants $M > m > 0$ such that

$$m \le \int_{R^n} k(t,s)ds \le M, \quad \forall\ t \in R^n,$$

and

$$\lim_{t\to t_0} \int_{R^n} |k(t,s) - k(t_0,s)|ds = 0, \quad \forall\ t_0 \in R^n;$$

and (iii) $a_i(t)$ $(i = 1,2,3,\ldots)$ are measurable and nonnegative on R^n and there exist constants $\sigma > \tau > 0$ such that

$$\tau \le \sum_{i=1}^{\infty} a_i(t) \le \sigma, \quad \forall\ t \in R^n.$$

Then, for any $\lambda > 0$ equation (2.2.62) (and also equation (2.2.63)) has exactly one continuous solution $x_\lambda(t)$ satisfying

$$0 < \inf_{t\in R^n} x_\lambda(t) \le \sup_{t\in R^n} x_\lambda(t) < +\infty \,.$$

Moreover, we have

(a) $0 < \lambda_1 < \lambda_2$ implies $\inf_{t\in R^n} [x_{\lambda_1}(t) - x_{\lambda_2}(t)] > 0$;

(b) $\sup_{t\in R^n} |x_\lambda(t) - x_{\lambda_0}(t)| \to 0$ as $\lambda \to \lambda_0$, where $\lambda_0 > 0$;

and

(c) $\sup_{t\in R^n} |x_\lambda(t)| \to 0$ as $\lambda \to +\infty$ and $\sup_{t\in R^n} |x_\lambda(t)| \to +\infty$ as $\lambda \to +0$.

Proof. Let E be the Banach space of all bounded and continuous functions $x(t)$ on R^n with norm

$$\|x\| = \sup_{t\in R^n} |x(t)|$$

and $P = \{x(t) \in E \mid x(t) \geq 0, \ \forall\, t \in R^n\}$. Obviously, P is a solid normal cone of E and $\overset{\circ}{P} = \{x(t) \in E \mid \inf_{t \in R^n} x(t) > 0\}$. Define

$$A_1 x(t) = \int_{R^n} k(t,x) \cdot \left\{ \sum_{i=1}^{\infty} a_i(s)\, x(s)^{\alpha_i} \right\} ds$$

and

$$A_2 x(t) = \int_{R^n} k(t,s) \cdot \left\{ \sum_{i=1}^{\infty} a_i(s)\, x(s)^{\alpha_i} \right\}^{-1} ds \, .$$

It is not difficult to verify that $A_1 : \overset{\circ}{P} \to \overset{\circ}{P}$ is α-concave and increasing and $A_2 : \overset{\circ}{P} \to \overset{\circ}{P}$ is $(-\alpha)$-convex and decreasing, where $\alpha = \sup_i \alpha_i$, $0 < \alpha < 1$. Hence, the existence and uniqueness of the solutions to equations (2.2.62) and (2.2.63) and the conclusions (a), (b), and (c) can be deduced from Theorem 2.2.6 and Theorem 2.2.7. □

2.3 *Fixed Points of Cone Expansion and Compression.*

(a) Fixed Point Index.

Let E be a real Banach space. A subset $X \subset E$ is called a retract of E if there exists a continuous mapping $r : E \to X$, and a retraction, when $r(x) = x$, $x \in X$. By a theorem due to Dugundji (see Dugundji [1]), every nonempty closed convex subset of E is a retract of E. In particular, every cone of E is a retract of E.

Theorem 2.3.1. Let X be a retract of real Banach space E. Then, for every relatively bounded open subset U of X and every completely continuous operator $A : \overline{U} \to X$ which has no fixed points on ∂U, there exists an integer $i(A,U,X)$ satisfying the following conditions:

(i) Normality: $i(A,U,X) = 1$ if $Ax \equiv y_0 \in U$ for any $x \in \overline{U}$.

(ii) Additivity: $i(A,U,X) = i(A,U_1,X) + i(A,U_2,X)$ whenever U_1 and U_2 are disjoint open subsets of U such that A has no fixed points on $\overline{U} \setminus (U_1 \cup U_2)$.

(iii) Homotopy invariance: $i(H(t,\cdot),U,X)$ is independent of t $(0 \leq t \leq 1)$ whenever $H : [0,1] \times \overline{U} \to X$ is completely continuous and $H(t,x) \neq x$ for any $(t,x) \in [0,1] \times \partial U$.

(iv) Permanence: $i(A,U,X) = i(A,U \cap Y,Y)$ if Y is a retract of X and $A(\overline{U}) \subset Y$.

Moreover, let

$$M = \{(A,U,X) \mid X \text{ retract of } E, \ U \text{ bounded open in } X, \\ A : \overline{U} \to X \text{ completely continuous and } Ax \neq x \text{ on } \partial U\}$$

and let Z be the set of integers. Then there exists exactly one function $d : M \to Z$ satisfying (i)-(iv). In other words, $i(A,U,X)$ is uniquely defined. $i(A,U,X)$ is called the <u>fixed point index</u> of A on U with respect to X.

<u>Proof</u>. First we prove the uniqueness of the fixed point index. Let $\{i(A,U,X)\}$ be any family satisfying conditions (i)-(iv). We define

$$d(f,U,p) = i(A + p,U,E), \tag{2.3.1}$$

where $f = I - A$, U bounded open set of E, $f(x) \neq p$ on ∂U, i.e., $A + p$ has no fixed points on $\partial\Omega$. From the conditions (i)-(iv) and (2.3.1) it is easy to show that the function $d(f,U,p)$ has the four properties which characterize the Leray-Schauder degree (see Appendix, Theorem A.3.2) and hence, by the uniqueness of Leray-Schauder degree, we have

$$d(f,U,p) = \deg(I - A,U,p). \tag{2.3.2}$$

Taking $p = \theta$ in (2.3.1) and (2.3.2), we get

$$i(A,U,E) = \deg(I - A,U,\theta). \tag{2.3.3}$$

Suppose now that X is an arbitrary retract of E and denote by $r : E \to X$ an arbitrary retraction. For open subset U of X, we choose a ball $B_R = \{x \in E \mid \|x\| < R\}$ such that $B_R \supset U$. Then, by the Permanence property (iv) and (2.3.3), we have

$$i(A,U,X) = i(A\cdot r, B_R \cap r^{-1}(U), E) = deg(I - A\cdot r, B_R \cap r^{-1}(U), \theta). \quad (2.3.4)$$

Hence, (2.3.4) and the uniqueness of the Leray-Schauder degree imply the uniqueness of the fixed point index.

By the above uniqueness proof we are led to define

$$i(A,U,X) = \deg(I - A\cdot r, B_R \cap r^{-1}(U), \theta), \quad (2.3.5)$$

where $r : E \to X$ is an arbitrary retraction and $B_R = \{x \in E \mid \|x\| < R\} \supset U$. Evidently, $B_R \cap r^{-1}(U)$ is a bounded open set of E and

$$\overline{B_R \cap r^{-1}(U)} \subset \overline{r^{-1}(U)} \subset r^{-1}(\overline{U}) . \quad (2.3.6)$$

It is easy to see that

$$x_0 \in r^{-1}(\overline{U}),\ A\cdot r(x_0) = x_0 \Rightarrow x_0 \in U,\ Ax_0 = x_0 . \quad (2.3.7)$$

Now, we prove that $i(A,U,X)$ defined by (2.3.5) is independent of the choice of R and r. Let $R_1 > R$. Since

$$U \subset B_R \cap r^{-1}(U) \subset B_{R_1} \cap r^{-1}(U),$$

by (2.3.7) we know that $A\cdot r$ has no fixed points in $\overline{B_{R_1} \cap r^{-1}(U)} \setminus (B_R \cap r^{-1}(U))$, and consequently, by the excision property of Leray-Schauder degree (see Appendix, Theorem A.3.3),

$$\deg(I - A\cdot r, B_R \cap r^{-} (U), \theta) = \deg(I - A\cdot r, B_R \cap r (U), \theta),$$

i.e., $i(A,U,X)$ is independent of the choice of R. Next, let $r_1 : E \to X$ be another retraction of E and let $V = B_R \cap r^{-1}(U) \cap r_1^{-1}(U)$. Then V is a bounded open set of E and $V \supset U$. By (2.3.7) we know that $A \cdot r$ has no fixed points in $\overline{B_R \cap r^{-1}(U)} \setminus V$ and $A \cdot r_1$ has no fixed points in $\overline{B_R \cap r_1^{-1}(U)} \setminus V$. Hence

$$\deg(I - A \cdot r, B_R \cap r^{-1}(U), \theta) = \deg(I - A \cdot r, V, \theta) \tag{2.3.8}$$

and

$$\deg(I - A \cdot r_1, B_R \cap r_1^{-1}(U), \;) = \deg(I - A \cdot r_1, V, \theta). \tag{2.3.9}$$

Now, let $h(t,x) = x - H(t,x)$, where $H(t,x) = r[tA \cdot r(x) + (1 - t)A \cdot r_1(x)]$. Clearly, $H : [0,1] \times \overline{V} \to E$ is completely continuous. We now prove $\theta \notin h(t, \partial V)$ for any $t \in [0,1]$. In fact, if there exist $t_0 \in [0,1]$ and $x_0 \in \partial V$ such that $h(t_0, x_0) = \theta$, then

$$x_0 = r[t_0 A \cdot r(x_0) + (1 - t_0)A \cdot r_1(x_0)] \in X.$$

As a result, $r(x_0) = x_0$, $r_1(x_0) = x_0$ and $x_0 = Ax_0$. And so, by (2.3.7), $x_0 \in U \subset V$, in contradiction with $x_0 \in \partial V$. Thus, using the homotopy invariance property of the Leray-Schauder degree and observing that $H(0,x) = r[A \cdot r_1(x)] = A \cdot r_1(x)$ and $H(1,x) = r[A \cdot r(x)] = A \cdot r(x)$, we have

$$\deg(I - A \cdot r_1, V, \theta) = \deg(I - A \cdot r, V, \theta). \tag{2.3.10}$$

It follows from (2.3.8), (2.3.9), and (2.3.10) that

$$\deg(I - A \cdot r, B_R \cap r^{-1}(U), \theta) = \deg(I - A \cdot r_1, B_R \cap r_1^{-1}(U), \theta), \tag{2.3.11}$$

which shows that $i(A,U,X)$ is independent of the choice of r.

Finally, by the basic properties of the Leray-Schauder degree (see Appendix, Theorems A.3.2 and A.3.3), it is not difficult to verify that the fixed point index defined by (2.3.5) has the properties (i)-(iv). We omit the details which can be found in Guo [1].

Theorem 2.3.2. Bedised (i)-(iv), the fixed point index has the following properties:

(v) Excision property: $i(A,U,X) = i(A,U_0,X)$ whenever U_0 is an open subset of U such that A has no fixed points in $\overline{U} \setminus U_0$.

(vi) Solution property: if $i(A,U,X) \neq 0$, then A has at least one fixed point in U.

Proof. Let $U_1 = U$ and $U_2 = \phi$ in additivity property (ii); we get $i(A,\phi,X) = 0$. From this and setting $U_1 = U_0$ and $U_2 = \phi$ in (ii), we obtain $i(A,U,X) = i(A,U_0,X)$. Thus, (v) is proved.

If A has no fixed points in U, letting $U_0 = \phi$ in (v), we get

$$i(A,U,X) = i(A,\phi,X) = 0,$$

and hence (vi) is proved. □

Notice that the concept of fixed point index can be extended to strict-set-contractions and condensing mappings, as follows.

Let X be a nonempty convex closed set of the real Banach space E and U be an open subset of X. Let operator $A : \overline{U} \to X$ be a k-set-contraction $(0 \leq k < 1)$, which has no fixed points on ∂U. Let

$$D_1 = \overline{co}A(\overline{U}), \quad D_n = \overline{co}A(D_{n-1} \cap \overline{U}) \quad (n = 2,3,4,\ldots). \tag{2.3.12}$$

If $D_n \cap \overline{U} = \phi$ for some n, we define the fixed point index

$$i(A,U,X) = 0. \tag{2.3.13}$$

Now, suppose $D_n \cap \overline{U} \neq \phi$ for any n. Evidently, $D_n \subset X$ $(n = 1,2,3,\ldots)$. We can prove that the set $D = \bigcap_{n=1}^{\infty} D_n \subset X$ is nonempty, convex, and compact and $D \cap \overline{U}$ is nonempty and compact and, moreover, $A(D \cap \overline{U}) \subset D$ (see Appendix A.3(c)). Hence, by the extension theorem, there exists a completely continuous operator $A_1 : \overline{U} \to D$ $(D \subset X)$ such that $A_1 x = Ax$ for any $x \in D \cap \overline{U}$. It is easy to show that A_1 has no fixed points on ∂U and hence, by Theorem 2.3.1, the fixed point index $i(A_1,U,X)$ is well defined. Now, we define

$$i(A,U,X) = i(A_1,U,X). \tag{2.3.14}$$

It is not difficult to prove that the fixed point index for k-set-contraction $(0 \leq k < 1)$ $i(A,U,X)$ defined by (2.3.13) and (2.3.14) is independent of the choice of A_1, and that Theorems 2.3.1 and 2.3.2 remain true.

Now, let the nonempty convex closed set X be starred, i.e., $x \in X$ implies $tx \in X$ for any $t \in [0,1]$. Let U be an open subset of X and $A : \overline{U} \to X$ be condensing without fixed points on ∂U. Choose a strict-set-contraction $B : \overline{U} \to X$ such that

$$\|Ax - Bx\| < \tau, \quad \forall\ x \in \overline{U},$$

where

$$\tau = \inf_{x \in \partial U} \|x - Ax\| > 0 .$$

Such B exists, for example, choose $B = kA$, where $0 \leq k < 1$ and $1 - k$ is sufficiently small. Since X is starred, B maps $\overline{U}$ into X. It is easy to see that B has no fixed points on ∂U and hence $i(B,U,X)$ is well defined. Now, we define

$$i(A,U,X) = i(B,U,X) .$$

It is easy to show that the fixed point index for condensing mappings defined by (2.3.15) is independent of the choice of B and for which Theorems 2.3.1 and 2.3.2 are true.

(b) Fixed Point Theorems of Cone Expansion and Compression.

In the following, let P be a cone of real Banach space E. Hence, P is a retract of E, and also P is a starred convex closed set. Let Ω be a bounded open set of E, then $P \cap \Omega$ is a bounded open set of P and $\partial(P \cap \Omega) = P \cap \partial\Omega$, $\overline{P \cap \Omega} = P \cap \overline{\Omega}$.

Lemma 2.3.1. Let $\theta \in \Omega$ and $A : P \cap \overline{\Omega} \to P$ be condensing. Suppose that

$$Ax \neq \mu x, \quad \forall x \in P \cap \Omega\partial, \quad \mu \geq 1. \tag{2.3.16}$$

Then $i(A,P \cap \Omega,P) = 1$.

Proof. Let $H(t,x) = tAx$. Then $H : [0,1] \times (P \cap \overline{\Omega}) \to P$ is continuous, and the continuity of $H(t,x)$ in t is uniform with respect to $x \in P \cap \overline{\Omega}$. Evidently, $H(t,\cdot) : P \cap \overline{\Omega} \to P$ is condensing for every $t \in [0,1]$ and $H(t,x) \neq x$ for $x \in P \cap \partial\Omega$ and $0 \leq t \leq 1$. Hence, by the homotopy invariance and normality of fixed point index, we have

$$i(A,P \cap \Omega,P) = i(\theta,P \cap \Omega,P) = 1 . \square$$

Lemma 2.3.2. Let $A : P \cap \overline{\Omega} \to P$ be completely continuous and $B : P \cap \partial\Omega \to P$ be completely continuous. Suppose that

(a) $\inf_{x \in P \cap \partial\Omega} \|Bx\| > 0;$

and

(b) $x - Ax \neq tBx, \quad \forall x \in P \cap \partial\Omega, \quad t \geq 0.$

Then, we have

$$i(A, P \cap \Omega, P) = 0. \tag{2.3.17}$$

Proof. By the extension theorem of Dugundji [1], we can extend B to a completely continuous operator from $P \cap \bar{\Omega}$ into P such that

$$B(P \cap \bar{\Omega}) \subset \overline{co}\, B(P \cap \partial\Omega) . \tag{2.3.18}$$

Let $F = B(P \cap \partial\Omega)$, then $\overline{co}\, B(P \cap \partial\Omega) = \overline{co}\, F = \bar{M}$, where

$$M = \left\{ y = \sum_{i=1}^{n} \lambda_i y_i \mid y_i \in F, \quad \lambda_i \geq 0, \quad \sum_{i=1}^{n} \lambda_i = 1; \quad n = 1,2,3,\ldots \right\}.$$

We first prove

$$\inf_{y \in \bar{M}} \|y\| > 0. \tag{2.3.19}$$

Denote by E_0 the subspace of E spanned by F. Since B is completely continuous, F is relatively compact, therefore E_0 is separable. Evidently, $P_0 = P \cap E_0$ is a cone of E_0 and $F \subset P_0$, $\overline{co}\, F \subset P_0$. By virtue of Theorem 1.4.1, there exists $f_0 \in E_0^*$ such that $f_0(y) > 0$ for any $y \in P_0$ with $y \neq \theta$. We claim that

$$\inf_{y \in F} f_0(y) = \sigma > 0. \tag{2.3.20}$$

In fact, if $\sigma = 0$, then there exists $\{y_k\} \subset F$ such that $f_0(y_k) \to 0$. By the relative compactness of F, there is a subsequence $\{y_{k_i}\}$ of $\{y_k\}$ such that $y_{k_i} \to y_0 \in P_0$, and so $f_0(y_{k_i}) \to f_0(y_0)$ and $f_0(y_0) = 0$. Hence $y_0 = \theta$ and $\|y_{k_i}\| \to 0$, which contradicts hypothesis (a). Thus, (2.3.20) holds.

For any $y = \sum_{i=1}^{n} \lambda_i y_i \in M$, where $y_i \in F$, $\lambda_i \geq 0$ and $\sum_{i=1}^{n} \lambda_i = 1$, we have by (2.3.20)

$$f_0(y) = \sum_{i=1}^{n} \lambda_i f_0(y_i) \geq \sum_{i=1}^{n} \lambda_i \sigma = \sigma,$$

and therefore,

$$f_0(y) \geq \sigma, \ \forall \ y \in \overline{M} . \tag{2.3.21}$$

Since $\overline{M} = \overline{co}\ F$ is compact, there exists a $z_0 \in \overline{M}$ such that

$$\inf_{y \in \overline{M}} \|y\| = \|z_0\| . \tag{2.3.22}$$

By (2.3.21), $f_0(z_0) \geq \sigma$, and this implies that $z_0 \neq \theta$. It follows therefore from (2.3.22) that (2.3.19) holds. By (2.3.18) and (2.3.19), we get

$$\inf_{x \in P \cap \overline{\Omega}} \|Bx\| = a > 0 . \tag{2.3.23}$$

Now, it is easy to show that (2.3.17) holds. In fact, if $i(A, P \cap \Omega, P) \neq 0$, then by hypothesis (b) and the homotopy invariance property of fixed point index, we have

$$i(A + tB, P \cap \Omega, P) = i(A, P \cap \Omega, P) \neq 0, \ \forall \ t > 0 .$$

In particular, choosing $t_0 > \dfrac{b + c}{a}$, where $b = \sup_{x \in P \cap \overline{\Omega}} \|x\|$ and $c = \sup_{x \in P \cap \overline{\Omega}} \|Ax\|$, we have

$$i(A + t_0 B, P \cap \Omega, P) \neq 0,$$

and so, by the solution property of fixed point index, there exists an $x_0 \cap P \in \Omega$ such that $A_{x_0} + t_0 B x_0 = x_0$. Hence

$$t_0 = \frac{\|x_0 - Ax_0\|}{\|Bx_0\|} \leq \frac{b + c}{a}$$

which is a contradiction. □

<u>Corollary 2.3.1</u>. Let $A : P \cap \overline{\Omega} \to P$ be completely continuous. If there exists a $u_0 > \theta$ such that

$$x - Ax \neq tu_0, \quad \forall\, x \in P \cap \partial\Omega, \quad t \geq 0, \tag{2.3.24}$$

then (2.3.17) holds.

Proof. Corollary 2.3.1 follows directly from Lemma 2.3.2 by putting $Bx = u_0$ for any $x \in P \cap \partial\Omega$. □

Lemma 2.3.3. Let $A : P \cap \overline{\Omega} \to P$ be completely continuous. Suppose that

(i) $\inf\limits_{x \in P \cap \partial\Omega} \|Ax\| > 0$;

and

(ii) $Ax \neq \mu x, \quad \forall\, x \in P \cap \partial\Omega, \quad 0 < \mu \leq 1$.

Then (2.3.17) holds.

Proof. Taking $B = A$ in Lemma 2.3.2, we see that condition (a) of Lemma 2.3.2 is the same as condition (i) of Lemma 2.3.3. Also, condition (b) of Lemma 2.3.2 is true. In fact, if there exist $x_0 \in P \cap \partial\Omega$ and $t_0 \geq 0$ such that $x_0 - Ax_0 = t_0 Ax_0$, then $Ax_0 = \mu x_0$, where $\mu_0 = (1 + t_0)^{-1}$. Evidently $0 < \mu_0 \leq 1$, which contradicts the condition (ii). Thus, (2.3.17) follows from Lemma 2.3.2. □

Notice, Lemma 2.3.3 is not true for strict-set-contraction A. For example, let $E = \ell^2 = \left\{x = (x_1, x_2, \ldots, x_n, \ldots) \mid \|x\| = \left(\sum\limits_{n=1}^{\infty} x_n^2\right)^{1/2} < +\infty\right\}$, $P = \{x = (x_1, x_2, \ldots, x_n, \ldots) \in \ell^2 \mid x_n \geq 0,\ n = 1, 2, 3, \ldots\}$, then P is a cone of E. We define the operator A by

$$Ax = (0, \tfrac{1}{2} x_1, \tfrac{1}{2} x_2, \ldots, \tfrac{1}{2} x_{n-1}, \ldots),$$

$$\forall\, x = (x_1, x_2, \ldots, x_n, \ldots) \in \ell^2. \tag{2.3.25}$$

Let $\Omega = \{x = (x_1, x_2, \ldots, x_n, \ldots) \in \ell^2 \mid \|x\| < 1\}$. It is clear that $A : P \cap \overline{\Omega} \to P$ and, since $\|Ax\| = 1/2\, \|x\|$ and $\|Ax - Ay\| = 1/2\, \|x - y\|$ for any $x, y \in \ell^2$, A is a 1/2-set-contraction. It is easy to check

$$Ax \neq \mu x, \quad \forall\, x \in P \cap \partial\Omega, \quad \mu > 0. \tag{2.3.26}$$

In fact, if there exist $x^* = (x_1^*, x_2^*, \ldots, x_n^*, \ldots) \in P \cap \partial\Omega$ and $\mu^* > 0$ such that $Ax^* = \mu^* x^*$, then

$$0 = \mu^* x_1^*, \ \frac{1}{2} x_2^* = \mu^* x_2^*, \ldots, \ \frac{1}{2} x_{n-1}^* = \mu^* x_n^*, \ldots,$$

and hence $x_1^* = x_2^* = \ldots = x_n^* = \ldots = 0$, i.e., $x^* = \theta$, in contradiction with $\|x^*\| = 1$. From $\|Ax\| = 1/2 \|x\|$ and (2.3.26) we see that conditions (i) and (ii) of Lemma 2.3.3 are satisfied. But by (2.3.26) and Lemma 2.3.1, we have $i(A, P \cap \partial\Omega, P) = 1$.

Lemma 2.3.4. Let $A : P \cap \overline{\Omega} \to P$ be completely continuous. Suppose that

(i') $Ax \neq \mu x, \ \forall \ x \in P \cap \partial\Omega, \ 0 \leq \mu \leq 1,$

and

(ii') the set $\{\|Ax\|^{-1} Ax \mid x \in P \cap \partial\Omega\}$ is relatively compact.

Then (2.3.17) holds.

Proof. Let $A_1 x = \alpha(\|Ax\|)^{-1} Ax$, where $\alpha = \sup_{x \in P \cap \partial\Omega} \|Ax\| > 0$. Then, by hypotheses, $A_1 : P \cap \partial\Omega \to P$ is completely continuous. By the extension theorem, A_1 can be extended to a completely continuous operator from $P \cap \overline{\Omega}$ into P. We now prove that A_1 satisfies the conditions (i) and (ii) of Lemma 2.3.3. In fact, first we have

$$\inf_{x \in P \cap \partial\Omega} \|A_1 x\| = \alpha > 0.$$

Secondly, if there exist $x_0 \in P \cap \partial\Omega$ and $0 < \mu_0 \leq 1$ such that $A_1 x_0 = \mu_0 x_0$, then $Ax_0 = \lambda_0 x_0$, where $\lambda_0 = \mu_0 \alpha^{-1} \|Ax_0\|$. Evidently, $0 < \lambda_0 \leq \mu_0 \leq 1$, which contradicts hypothesis (i'). Hence, by Lemma 2.3.3, we have

$$i(A_1, P \cap \Omega, P) = 0. \tag{2.3.27}$$

Now, we prove

$$(1 - t)Ax + tA_1 x \neq x, \ \forall \ x \in P \cap \partial\Omega, \ 0 \leq t \leq 1. \tag{2.3.28}$$

If there are $x_1 \in P \cap \partial\Omega$ and $0 \leq t_1 \leq 1$ such that $(1 - t_1)Ax_1 + t_1A_1x_1 = x_1$, then $Ax_1 = \mu_1x_1$, where $\mu_1 = [1 + t_1(\alpha/Ax_1 - 1)]^{-1}$, $0 \leq \mu_1 \leq 1$, in contradiction with hypothesis (i'). Hence, by (2.3.27), (2.3.28), and the homotopy invariance property of fixed point index, we get

$$i(A,P \cap \Omega,P) = i(A_1,P \cap \Omega,P) = 0. \ \square$$

Theorem 2.3.3. (Fixed point theorem of cone expansion and compression). Let Ω_1 and Ω_2 be two bounded open sets in E such that $\theta \in \Omega_1$ and $\overline{\Omega}_1 \subset \Omega_2$. Let operator $A : P \cap (\overline{\Omega}_2 \setminus \Omega_1) \to P$ be completely continuous. Suppose that one of the two conditions

$$(\mathrm{H}_1) \quad Ax \not\leq x, \quad \forall\, x \in P \cap \partial\Omega_1 \quad \text{and} \quad Ax \not\geq x, \quad \forall\, x \in P \cap \partial\Omega_2$$

and

$$(\mathrm{H}_2) \quad Ax \not\geq x, \quad \forall\, x \in P \cap \partial\Omega_1 \quad \text{and} \quad Ax \not\leq x, \quad \forall\, x \in P \cap \partial\Omega_2$$

is satisfied. Then A has at least one fixed point in $P \cap (\Omega_2 \setminus \overline{\Omega}_1)$.

Proof. By the extension theorem, A has a completely continuous extension (also denoted by A) from $P \cap \overline{\Omega}_2$ into P.

First we assume that (H_1) is satisfied, i.e., it is the case of cone expansion. It is easy to see that

$$Ax \neq \mu x, \quad \forall\, x \in P \cap \partial\Omega_1, \quad \mu \geq 1, \tag{2.3.29}$$

since, otherwise, there exist $x_0 \in P \cap \partial\Omega_1$ and $\mu_0 \geq 1$ such that $Ax_0 = \mu_0x_0 \geq x_0$, in contradiction with (H_1). Now, from (2.3.29) and Lemma 2.3.1, we obtain

$$i(A,P \cap \Omega_1,P) = 1. \tag{2.3.30}$$

On the other hand, choosing an arbitrary $u_0 > \theta$, we have

$$x - Ax \neq tu_0, \quad \forall\, x \in P \cap \partial\Omega_2, \quad t \geq 0. \tag{2.3.31}$$

In fact, if there exist $x_1 \in P \cap \partial\Omega_2$ and $t_1 \geq 0$ such that $x_1 - Ax_1 = t_1 u_0 \geq \theta$, then $x_1 \geq Ax_1$, in contradiction with (H_1). Hence, by (2.3.31) and Corollary 2.3.1, we have

$$i(A, P \cap \Omega_2, P) = 0. \tag{2.3.32}$$

It follows therefore from (2.3.30), (2.3.32) and the additivity property of fixed point index that

$$i(A, P \cap (\Omega_2 \setminus \overline{\Omega}_1), P) = i(A, P \cap \Omega_2, P) - i(A, P \cap \Omega_1, P) = -1 \neq 0. \tag{2.3.33}$$

Hence, by the solution property of fixed point index, A has at least one fixed point in $\Omega_2 \setminus \overline{\Omega}_1$.

Similarly, when (H_2) is satisfied, instead of (2.3.30), (2.3.32), and (2.3.33), we have $i(A, P \cap \Omega_1, P) = 0$, $i(A, P \cap \Omega_2, P) = 1$, and $i(A, P \cap (\Omega_2 \setminus \overline{\Omega}_1), P) = 1$. As a result we also can assert that A has at least one fixed point in $\Omega_2 \setminus \overline{\Omega}_1$. □

Theorem 2.3.4. (Fixed point theorem of cone expansion and compression of norm type). Let Ω_1 and Ω_2 be two bounded open sets in E such that $\theta \in \Omega_1$ and $\overline{\Omega}_1 \subset \Omega_2$. Let operator $A : P \cap (\overline{\Omega}_2 \setminus \Omega_1) \to P$ be completely continuous. Suppose that one of the two conditions

(H_3) $\|Ax\| \leq \|x\|$, $\forall x \in P \cap \partial\Omega_1$ and $\|Ax\| \geq \|x\|$, $\forall x \in P \cap \partial\Omega_2$

and

(H_4) $\|Ax\| \geq \|x\|$, $\forall x \in P \cap \partial\Omega_1$ and $\|Ax\| \leq \|x\|$, $\forall x \in P \cap \partial\Omega_2$

is satisfied. Then A has at least one fixed point in $P \cap (\overline{\Omega}_2 \setminus \Omega_1)$.

Proof. We only need to prove this theorem under condition (H_3), since the proof is similar when (H_4) is satisfied. By the extension theorem, A can be extended to a completely continuous operator from $P \cap \overline{\Omega}_2$ into P. We may assume that A has no fixed points on $P \cap \partial\Omega_1$ and $P \cap \partial\Omega_2$. It is easy to see that (2.3.29) holds, since otherwise, there exist $x_0 \in P \cap \partial\Omega_1$

and $\mu_0 > 1$ such that $Ax_0 = \mu_0 x_0$ and hence $\|Ax_0\| = \mu_0\|x_0\| > \|x_0\|$, in contradiction with (H_3). Thus, by (2.3.29) and Lemma 2.3.1, (2.3.30) holds.

On the other hand, it is also easy to verify

$$Ax \neq \mu x, \ \forall \ x \in P \cap \partial\Omega_2, \quad 0 < \mu \leq 1. \tag{2.3.34}$$

In fact, if there are $x_1 \cap P \in \partial\Omega_2$ and $0 < \mu_1 < 1$ such that $Ax_1 = \mu_1 x_1$, then $\|Ax_1\| = \mu_1\|x_1\| < \|x_1\|$, in contradiction with (H_3). In addition, by (H_3) we have

$$\inf_{x\in P\cap\partial\Omega_2} \|Ax\| \geq \inf_{x\in P\cap\partial\Omega_2} \|x\| > 0. \tag{2.3.35}$$

It follows from (2.3.34), (2.3.35), and Lemma 2.3.3 that (2.3.32) holds. As before, (2.3.30) and (2.3.32) imply (2.3.33), and therefore A has at least one fixed point in $\Omega_2 \setminus \overline{\Omega}_1$. □

Theorem 2.3.5. Let Ω_1 and Ω_2 be two bounded open sets in E such that $\theta \in \Omega_1$ and $\overline{\Omega}_1 \subset \Omega_2$. Let $u_0 > \theta$, $P_{u_0} = \{x \in P \mid \exists \lambda > 0, \ x \geq \lambda u_0\}$, and $A : P \cap (\overline{\Omega}_2 \setminus \Omega_1) \to P$ be completely continuous. If one of the two conditions

(H_5) $Ax \not\geq (1 + \varepsilon)x, \ \forall x \in P \cap \partial\Omega_1, \ \varepsilon > 0$ and $Ax \not\leq x$, $\forall x \in P_{u_0} \cap \partial\Omega_2$

and

(H_6) $Ax \not\leq x, \ \forall \ x \in P_{u_0} \cap \partial\Omega_1$ and $Ax \not\geq (1 + \varepsilon)x$, $\forall \ x \in P \cap \partial\Omega_2, \ \varepsilon > 0$

is satisfied, then A has at least one fixed point in $P \cap (\overline{\Omega}_2 \setminus \Omega_1)$.

Proof. We may assume that (H_5) is satisfied, because the proof is similar when (H_6) holds. By the extension theorem, A can be extended to a completely continuous operator from $P \cap \overline{\Omega}_2$ into P. We may assume that A

has no fixed points on $P \cap \partial\Omega_1$ and $P \cap \partial\Omega_2$. It is easy to show that (2.3.29) holds, since otherwise, there are $x_0 \in P \cap \partial\Omega_1$ and $\mu_0 > 1$ such that $Ax_0 = \mu_0 x_0 = (1 + \varepsilon_0)x_0$, where $\varepsilon_0 = \mu_0 - 1 > 0$, which contradicts (H_5). Hence (2.3.30) holds.

On the other hand, (2.3.31) must also hold. In fact, if there exist $x_1 \in P \cap \partial\Omega_2$ and $t_1 > 0$ such that $x_1 - Ax_1 = t_1 u_0$, then $x_1 \geq t_1 u_0$ and $x_1 \geq Ax_1$, and so $x_1 \in P_{u_0}$, in contradiction with (H_5). Hence (2.3.32) is true.

It follows from (2.3.30) and (2.3.32) that (2.3.33) is also true and therefore, A has at least one fixed point in $P \cap (\Omega_2 \setminus \overline{\Omega}_1)$. □

Notice that since $P_{u_0} \subset P$, Theorem 2.3.5 is an improvement of Theorem 2.3.3.

Example 2.3.1. Consider the two-point boundary value problem of an ordinary differential equation:

$$-x'' = f(x), \quad 0 \leq t \leq 1; \quad x(0) = x(1) = 0, \tag{2.3.36}$$

where $f(x)$ is continuous and nonnegative for $x \geq 0$ and $f(0) = 0$. Obviously, $x(t) \equiv 0$ is the trivial solution of problem (2.3.36). We have the following conclusion: if

$$0 \leq \overline{\lim_{x\to+0}} \frac{f(x)}{x} < 8 \quad \text{and} \quad 24\sqrt{3} < \overline{\lim_{x\to+\infty}} \frac{f(x)}{x} \leq +\infty, \tag{2.3.37}$$

then the problem (2.3.36) has at least one nontrivial solution $x(t)$, which belongs to $C^2[0,1]$ and satisfies

$$x(t) > 0, \quad \forall\, 0 < t < 1. \tag{2.3.38}$$

To prove the conclusion, let us first observe that the problem (2.3.36) is equivalent to the integral equation

$$x(t) = \int_0^1 k(t,s)f[x(s)]ds \ ,$$

where $k(t,s)$ is the Green function of the differential operator $-x''$ with boundary condition $x(0) = x(1) = 0$; given by

$$k(t,s) = \begin{cases} t(1 - s), & t \le s; \\ s(1 - t), & t > s. \end{cases} \qquad (2.3.39)$$

Let

$$Ax(t) = \int_0^1 k(t,s)f[x(s)]ds \ ,$$

$P = \{x(t) \in C[0,1] \mid x(t) \ge 0\}$ and $P_\varepsilon = \{x(t) \in P \mid \min_{1/2-\varepsilon \le t \le 1/2-\varepsilon} x(t) \ge (1/2 - \varepsilon)\|x\|\}$, where $0 < \varepsilon < 1/2$ and $\|x\|$ denotes the norm of $x(t)$ in space $E = C[0,1]$. Evidently, P and P_ε are cones of $C[0,1]$, $P_\varepsilon \subset P$ and $A : P \to P$ is completely continuous. Let $x(t) \in P$. Since $k(t,s) \le s(1 - s)$,

$$\|Ax\| \le \int_0^1 s(1 - s)f[x(s)]ds \ . \qquad (2.3.40)$$

On the other hand, when $1/2 - \varepsilon \le t \le 1/2 + \varepsilon$, we have

$$k(t,s) = \begin{cases} t(1 - s) \ge (\frac{1}{2} - \varepsilon)(1 - s) & \text{for } t \le s; \\ s(1 - t) \ge s[1 - (\frac{1}{2} + \varepsilon)] = (\frac{1}{2} - \varepsilon) & \text{for } t > s, \end{cases}$$

and so

$$k(t,s) \ge (\frac{1}{2} - \varepsilon)s(1 - s), \ \forall \ \frac{1}{2} - \varepsilon \le t \le \frac{1}{2} + \varepsilon, \quad 0 \le s \le 1.$$

Hence

$$\min_{\frac{1}{2}-\varepsilon\le t\le\frac{1}{2}+\varepsilon} Ax(t) \ge (\frac{1}{2} - \varepsilon) \int_0^1 s(1 - s)f[x(s)]ds \ . \tag{2.3.41}$$

It follows from (2.3.40) and (2.3.41) that

$$\min_{\frac{1}{2}-\varepsilon\le t\le\frac{1}{2}+\varepsilon} Ax(t) \ge (\frac{1}{2} - \varepsilon)\|Ax\| \ ,$$

i.e., $Ax(t) \subset P_\varepsilon$. Thus $A(P) \subset P_\varepsilon$, and so

$$A(P_\varepsilon) \subset P_\varepsilon, \ \forall \ 0 < \varepsilon < \frac{1}{2} \ . \tag{2.3.42}$$

Now, by (2.3.37), there exist $0 < r < \eta < +\infty$ such that $0 \le f(x) \le 8x$ for $0 \le x \le r$ and $f(x) \ge 24\sqrt{3}x$ for $x \ge \eta$. Thus, for $x(t) \in P$ with $\|x\| = r$, we have

$$Ax(t) \le 8 \int_0^1 k(t,x)x(s)ds \le 8\|x\| \int_0^1 k(t,s)ds$$

$$= 4t(1 - t)\|x\| \le \|x\|, \ \forall \ 0 \le t \le 1,$$

and hence

$$\|Ax\| \le \|x\|, \ \forall \ x(t) \in P, \ \|x\| = r. \tag{2.3.43}$$

On the other hand, letting

$$R_\varepsilon = \max\left\{2r, \eta\left(\frac{1}{2} - \varepsilon\right)^{-1}\right\}, \ \forall \ 0 < \varepsilon < \frac{1}{2} \ , \tag{2.3.44}$$

we have $R_\varepsilon > r$ and, for $x(t) \in P_\varepsilon$ with $\|x\| = R_\varepsilon$,

$$\min_{\frac{1}{2}-\varepsilon\le t\le\frac{1}{2}+\varepsilon} x(t) \ge \left(\frac{1}{2} - \varepsilon\right)\|x\| = \left(\frac{1}{2} - \varepsilon\right)R_\varepsilon \ge \eta.$$

Consequently,

$$Ax\left(\frac{1}{2}\right) \ge \int_{\frac{1}{2}-\varepsilon}^{\frac{1}{2}+\varepsilon} k\left(\frac{1}{2},s\right)f[x(s)]ds \ge 24\sqrt{3} \int_{\frac{1}{2}-\varepsilon}^{\frac{1}{2}+\varepsilon} k\left(\frac{1}{2},s\right)x(s)ds$$

$$\geq 24\sqrt{3}\left(\frac{1}{2} - \varepsilon\right)\|x\| \int_{\frac{1}{2}-\varepsilon}^{\frac{1}{2}+\varepsilon} k\left(\frac{1}{2},s\right)ds = 12\sqrt{3}\left(\frac{1}{2} - \varepsilon\right)\varepsilon(1 - \varepsilon)\|x\|,$$

and therefore

$$\|Ax\| \geq 12\sqrt{3}\ \varepsilon(1 - \varepsilon)\left(\frac{1}{2} - \varepsilon\right)\|x\|, \ \forall \ x(t) \in P_\varepsilon, \ \|x\| = R_\varepsilon. \tag{2.3.45}$$

It is easy to see that the function $\psi(\varepsilon) = \varepsilon(1 - \varepsilon)(1/2 - \varepsilon)$ with $0 < \varepsilon < 1/2$ takes its maximum value at $\varepsilon = \varepsilon_0 = 3 - \sqrt{3}/6$ and $\psi(\varepsilon_0) = \sqrt{3}/36$. As a result, choosing $\varepsilon = \varepsilon_0$ in (2.3.45), we get

$$\|Ax\| \geq |x|, \ \forall \ x(t) \in P_{\varepsilon_0}, \ \|x\| = R_{\varepsilon_0}. \tag{2.3.46}$$

Now, observing (2.3.42), (2.3.43), and (2.3.46) and using Theorem 2.3.4 with $\Omega_1 = \{x(t) \in C[0,1] \mid \|x\| < r\}$, $\Omega_2 = \{x(t) \in C[0,1] \mid \|x\| < R_{\varepsilon_0}\}$ and P_{ε_0}, we conclude that A has a fixed point $x(t)$ in $P_{\varepsilon_0} \cap (\overline{\Omega}_2 \setminus \Omega_1)$. □

<u>Theorem 2.3.6</u>. Let Ω be a bounded open set in E and $\theta \in \Omega$. Suppose that $A : P \cap \overline{\Omega} \to P$ is completely continuous, $A\theta = \theta$ and

$$\inf_{x \in P \cap \partial\Omega} \|Ax\| > 0. \tag{2.3.47}$$

Then A has at least one eigenvector on $P \cap \partial\Omega$, which corresponds to positive eigenvalue, i.e., there exist $x_0 \in P \cap \partial\Omega$ and $\mu_0 > 0$ such that $Ax_0 = \mu_0 x_0$.

<u>Proof</u>. Let $m = \sup_{x \in P \cap \partial\Omega} \|x\|$, $\beta = \inf_{x \in P \cap \partial\Omega} \|Ax\|$. By (2.3.47), $m > 0$ and $\beta > 0$. Now, choose $a > m/\beta$ and it is easy to verify that operator aA satisfies all conditions of Lemma 2.3.3. In fact, if there exist $x_1 \in P \cap \partial\Omega$ and $0 < \mu_1 \leq 1$ such that $aAx_1 = \mu_1 x_1$, then

$$\mu_1\|x_1\| = a\|Ax_1\| > \frac{m}{\beta} \cdot \beta = m \geq \|x_1\|,$$

and hence $\mu_1 > 1$, a contradiction. Thus, by Lemma 2.3.3, we have

$$i(aA, P \cap \Omega, P) = 0. \tag{2.3.48}$$

Let $H(t,x) = taAx$. If $H(t,x) \neq x$ for any $x \in P \cap \partial\Omega$ and $0 \leq t \leq 1$, then, by the homotopy invariance property and normality property of fixed point index,

$$i(aA, P \cap \Omega, P) = i(\theta, P \cap \Omega, P) = 1,$$

in contradiction with (2.3.48). Hence, there exist $x_0 \in P \cap \partial\Omega$ and $0 \leq t_0 \leq 1$ such that $H(t_0, x_0) = x_0$, i.e., $t_0\, aAx_0 = x_0$. Obviously, $t_0 \neq 0$, and so $Ax_0 = \mu_0 x_0$, where $\mu_0 = (at_0)^{-1} > 0$. □

Theorem 2.3.6 can be extended to the case of k-set-contractions in some sense (see Massabo and Stuart [1]).

Example 2.3.2. Consider the eigenvalue problem corresponding to the problem (2.3.36):

$$\mu x'' + f(x) = 0, \quad 0 \leq t \leq 1; \quad x(0) = x(1) = 0\ , \tag{2.3.49}$$

where, as before, $f(x)$ is continuous and nonnegative for $x \geq 0$ and $f(0) = 0$. We have the following conclusion: if

$$0 < \lim_{x \to +\infty} \frac{f(x)}{x} \leq +\infty\ , \tag{2.3.50}$$

then there exists $R > 0$ such that, for every $r > R$, problem (2.3.49) has solution $x_r(t) \in C^2[0,1]$, $\mu_r > 0$ satisfying $x_r(t) > 0$ ($\forall\ 0 < t < 1$) and $\|x_r\| = \max_{0 \leq t \leq 1} x_r(t) = r$.

As in Example 2.3.1, problem (2.3.49) is equivalent to the integral equation

$$\mu x(t) = \int_0^1 k(t,x) f[x(s)]ds = Ax(t), \tag{2.3.51}$$

where $k(t,s)$ is the corresponding Green function (2.3.39). By the same method as in Example 2.3.1, we can prove that (2.3.42) holds. By (2.3.50), there exist $\tau > 0$ and $\eta > 0$ such that $f(x) > \tau x$ for $x \geq \eta$. Now, we fix ε $(0 < \varepsilon < 1/2)$ and prove that $R = \eta(1/2 - \varepsilon)^{-1}$ is required.

For $r > R$ we put $\Omega_r = \{x(t) \in C[0,1] \mid \|x\| < r\}$. Thus, when $x(t) \in P_\varepsilon \cap \partial\Omega_r$ we have

$$\min_{\frac{1}{2}-\varepsilon \leq t \leq \frac{1}{2}+\varepsilon} x(t) \geq \left(\frac{1}{2} - \varepsilon\right)\|x_r\| = \left(\frac{1}{2} - \varepsilon\right)r > \eta .$$

Consequently,

$$Ax\left(\frac{1}{2}\right) \geq \int_{\frac{1}{2}-\varepsilon}^{\frac{1}{2}+\varepsilon} k\left(\frac{1}{2},s\right)f[x(s)]ds \geq \tau \int_{\frac{1}{2}-\varepsilon}^{\frac{1}{2}+\varepsilon} k\left(\frac{1}{2},s\right)x(s)ds$$

$$\geq \tau\left(\frac{1}{2} - \varepsilon\right)\|x\| \int_{\frac{1}{2}-\varepsilon}^{\frac{1}{2}+\varepsilon} k\left(\frac{1}{2},s\right)ds = \frac{1}{2}\tau\,\varepsilon(1 - \quad)\left(\frac{1}{2} - \varepsilon\right) r,$$

and therefore

$$\inf_{x(t)\in P_\varepsilon \cap \partial\Omega_r} \|Ax\| \geq \frac{1}{2}\tau\,\varepsilon(1 - \varepsilon)\left(\frac{1}{2} - \varepsilon\right)r > 0 . \tag{2.3.52}$$

It follows from (2.3.52) and Theorem 2.3.6 that there exist $x(t) \in P_\varepsilon \cap \partial\Omega_r$ and $\mu_r > 0$ such that $Ax_r(t) = \mu_r x_r(t)$. □

Theorem 2.3.7. Let operator $A : P \to P$ be completely continuous and $A\theta = \theta$. Suppose that one of the two conditions

$$(H_7) \qquad \lim_{x\in P, \|x\|\to 0} \frac{\|Ax\|}{\|x\|} = 0, \qquad \lim_{x\in P, \|x\|\to +\infty} \frac{\|Ax\|}{\|x\|} = +\infty$$

and

$$(H_8) \qquad \lim_{x\in P, \|x\|\to 0} \frac{\|Ax\|}{\|x\|} = +\infty, \qquad \lim_{x\in P, \|x\|\to +\infty} \frac{\|Ax\|}{\|x\|} = 0$$

is satisfied. Then the following two conclusions hold:

(a) every $\mu > 0$ is an eigenvalue of A, which corresponds to positive eigenvector, i.e., there exists $x_\mu > \theta$ such that $Ax_\mu = \mu x_\mu$;

(b) $\lim_{\mu\to+\infty} \|x_\mu\| = +\infty$ under (H_7) and $\lim_{\mu\to+\infty} \|x_\mu\| = 0$ under (H_8).

Proof. We need only prove this theorem under condition (H_7) since the proof is similar when (H_8) is satisfied. For given $\mu > 0$, by virtue of (H_7), there exist $R > r > 0$ such that

$$\left\|\frac{1}{\mu} Ax\right\| < \|x\|, \quad \forall\ x \in P, \quad \|x\| = r$$

and

$$\left\|\frac{1}{\mu} Ax\right\| > \|x\|, \quad \forall\ x \in P, \quad \|x\| = R.$$

Hence, condition (H_3) of Theorem 2.3.4 is satisfied for operator $1/\mu\ A$ and $\Omega_1 = \{x \in E \mid \|x\| < r\}$, $\Omega_2 = \{x \in E \mid \|x\| < R\}$. It follows therefore from Theorem 2.3.4 that operator $1/\mu\ A$ has a fixed point x_μ in $\overline{\Omega}_2 \setminus \Omega_1$ and this proves the conclusion (a).

It remains to prove $\|x_\mu\| \to +\infty$ as $\mu \to +\infty$, i.e., conclusion (b). Suppose that this is not true. Then there exist a number $c > 0$ and a sequence $\mu_n \to +\infty$ such that

$$\|x_{\mu_n}\| \leq c \quad (n = 1,2,3,\ldots).$$

Furthermore, the sequence $\{\|x_{\mu_n}\|\}$ contains a subsequence that converges to a number τ $(0 \leq \tau \leq c)$. For simplicity, assume that $\{\|x_{\mu_n}\|\}$ itself converges to τ.

If $\tau > 0$, then $\|x_{\mu_n}\| > \tau/2$ for sufficiently large n (say, $n > N$), and hence

$$\mu_n = \frac{\|Ax_{\mu_n}\|}{\|x_{\mu_n}\|} \leq \frac{2M}{\tau} \quad (n > N),$$

where $M = \sup_{\|x\| \leq c} \|Ax\|$, which contradicts $\mu_n \to +\infty$.

If $\tau = 0$, then from (H_7) we have

$$\mu_n = \frac{\|Ax_{\mu_n}\|}{\|x_{\mu_n}\|} \to 0 \quad (n \to \infty),$$

in contradiction with $\mu_n \to +\infty$. Hence, $\|x_\mu\| \to +\infty$ as $\mu \to +\infty$ and our proof is complete.

(c) Fixed Point Theorems for Differentiable Operators.

Definition 2.3.1. Let P be a cone in real Banach space E and operator $A : P \to P$.

(i) Let $x_0 \in P$. If there exists a bounded linear operator $B : E \to E$ such that

$$\lim_{h \in P, \|h\| \to 0} \frac{\|A(x_0 + h) - Ax_0 - Bh\|}{\|h\|} = 0, \tag{2.3.53}$$

then A is said to be differentiable at x along the cone P, and B is called the Fréchet derivative of A at x_0 along P and is denoted by $A'_+(x_0)$, i.e., $A'_+(x_0) = B$.

(ii) If there exists a bounded linear operator $B_1 : E \to E$ such that

$$\lim_{x \in P, \|x\| \to +\infty} \frac{\|Ax - B_1 x\|}{\|x\|} = 0, \tag{2.3.54}$$

then A is said to be differentiable at infinity along the cone P, and B_1 is called the Fréchet derivative of A at infinity along P and is denoted by $A'_+(\infty)$, i.e., $A'_+(\infty) = B_1$.

Clearly, (2.3.53) is equivalent to

$$A(x_0 + h) - Ax_0 = Bh + \omega(x_0, h), \tag{2.3.55}$$

where $\|\omega(x_0,h)\| = o(\|h\|)$ as $h \to \theta$ $(h \geq \theta)$.

Example 2.3.3. Consider the Uryson integral operator

$$Ax(t) = \int_G k(t,s,x(s))ds, \tag{2.3.56}$$

where G is a bounded closed set in R^n, $k(t,s,x)$ is nonnegative and continuous for $t \in G$, $s \in G$, and $x \geq 0$. Let $E = C(G)$ and $P = \{x(t) \in C(G) \mid x(t) \geq 0,\ t \in G\}$, then P is a cone of E. It is obvious that $A : P \to P$ is completely continuous. We have the following conclusions:

(a) if $k_x'(t,s,x)$ exists and is continuous on $t \in G$, $s \in G$, and $x \geq 0$, then A is differentiable at any $x_0(t) \in P$ along P, and

$$A_+'(x_0)h(t) = \int_G k_x'(t,s,x_0(s))h(s)ds, \ \forall\ h(t) \in C(G); \tag{2.3.57}$$

(b) if

$$\frac{k(t,s,x)}{x} \to Q(t,s) \tag{2.3.58}$$

uniformly with respect to $t \in G$ and $s \in G$ as $x \to +\infty$, then A is differentiable at infinity along P, and

$$A_+'(\infty)h(t) = \int_G Q(t,s)h(s)ds, \ \forall\ h(t) \in C(G). \tag{2.3.59}$$

To prove the conclusion (a), we let

$$Bh(t) = \int_G k_x'(t,s,x_0(s))h(s)ds, \tag{2.3.60}$$

and observe that B is a completely continuous linear operator from $C(G)$ into $C(G)$. For $h(t) \in P$, we have

$$\begin{aligned} &|A[x_0(t) + h(t)] - Ax_0(t) - Bh(t)| \\ &\quad = \left|\int_G \{k(t,s,x_0(s) + h(s)) - k(t,s,x_0(s)) - k_x'(t,s,x_0(s))h(s)\}ds\right| \\ &\quad = \left|\int_G \{k_x'(t,s,x_0(s) + \theta(t,s)h(s)) - k_x'(t,s,x_0(s))\}h(s)ds\right| \\ &\quad \leq \|h\| \cdot \int_G |k_x'(t,s,x_0(s) + \theta(t,s)h(s)) - k_x'(t,s,x_0(s))|ds, \end{aligned} \tag{2.3.61}$$

where $0 < \theta(t,s) < 1$. For any given $\varepsilon > 0$, since $k'_x(t,s,x)$ is uniformly continuous on $G \times G \times [0,M+1]$, with $M = \max_{t\in G} x_0(t)$, there exists $0 < \delta < 1$ such that $t \in G$, $s \in G$ and x_1, $x_2 \in [0,M+1]$, $|x_1 - x_2| < \delta$ imply

$$|k'_x(t,s,x_1) = k'_x(t,s,x_2)| < \frac{\varepsilon}{\text{mes } G}. \tag{2.3.62}$$

It follows from (2.3.61) and (2.3.62) that

$$\|A(x_0 + h) - Ax_0 - Bh\| \leq \varepsilon\|h\|, \quad \forall\ h \in P, \quad \|h\| < \delta.$$

Hence (2.3.53) holds, and so $A'_+(x_0) = B$, ie., (2.3.57) is true.

To prove (b), for any given $\varepsilon > 0$, by hypothesis $Q(t,s)$ is continuous on $(t,s) \in G \times G$ and there exists $\beta > 0$ such that

$$|k(t,s,x) - Q(t,s)x| < \varepsilon\, x, \quad \forall\ t \in G, \quad s \in G, \quad x \geq \beta. \tag{2.3.63}$$

Let

$$\gamma = \max_{t\in G, s\in G, 0\leq x\leq\beta} |k(t,s,x) - Q(t,s)x|,$$

then, by (2.3.63), we have

$$|k(t,s,x) - Q(t,s)x| \leq \varepsilon\, x + \gamma, \quad \forall\ t \in G, \quad s \in G, \quad x \geq 0. \tag{2.3.64}$$

Now, define

$$B_1x(t) = \int_G Q(t,s)x(s)ds, \quad x(t) \in C(G).$$

Evidently, $B_1 : C(G) \to C(G)$ is a linear, completely continuous operator. From (2.3.64) we know, for any $x(t) \in P$,

$$|Ax(t) - B_1x(t)| = \left| \int_G \{k(t,s,x(s)) - Q(t,s)\}ds \right.$$
$$\leq \int_G \{\varepsilon x(s) + \gamma\}ds \leq \varepsilon(\text{mes } G)\|x\| + \gamma \text{ mes } G,$$

and hence

$$\overline{\lim_{x\in P, \|x\|\to+\infty}} \frac{\|Ax - B_1 x\|}{\|x\|} \leq \varepsilon \text{ (mes } G) .$$

Since ε is arbitrary, we see that (2.3.54) holds, and so $A'_+(\infty) = B_1$, i.e., (2.3.59), is true. □

Lemma 2.3.5. Let $A : P \to P$ be completely continuous. Then

(i) if A is differentiable at $x_0 \in P$ along P, then $A'_+(x_0)$ maps bounded sets of P into relatively compact sets;

(ii) if A is differentiable at infinite along P, then $A'_+(\infty)$ maps bounded sets of P into relatively compact sets.

Proof. (i) Since $A'_+(x_0)$ is linear, we need only prove that $A'_+(x_0)(S_1)$ is relatively compact, where $S_1 = \{x \in P \mid \|x\| \leq 1\}$. If otherwise, there exists $h_i \in S_1$ $(i = 1,2,3,\ldots)$ and $\varepsilon_0 > 0$ such that

$$\|A'_+(x_0)h_i - A'_+(x_0)h_j\| \geq \varepsilon_0 \quad (i \neq j). \tag{2.3.65}$$

Choose $\tau > 0$ such that

$$\|A(x_0 + h) - Ax_0 - A'_+(x_0)h\| \leq \frac{\varepsilon_0}{\varepsilon}\|h\|, \quad \forall\ h \in P, \quad \|h\| \leq \tau. \tag{2.3.66}$$

Then, from (2.3.65) and (2.3.66), we find

$$\begin{aligned}
\|A(x_0 + \tau h_i) - A(x_0 + \tau h_j)\| &= \| A(x_0 + \tau h_i) - Ax_0 - A'_+(x_0)(\tau h_i) \\
&\quad - A(x_0 + \tau h_j) - Ax_0 - A'_+(x_0)(\tau h_j) + \tau[A'_+(x_0)h_i = A'_+(x_0)h_j]\| \\
&\geq \tau\|A'_+(x_0)h_i - A'_+(x_0)h_j\| - \|A(x_0 + \tau h_i) - Ax_0 - A'_+(x_0)(\tau h_i)\| \\
&\quad - \|A(x_0 + \tau h_j) - Ax_0 - A'_+(x_0)(\tau h_j)\| \\
&\geq \tau\varepsilon_0 - \frac{\tau\varepsilon_0}{3} - \frac{\tau\varepsilon_0}{3} = \frac{\tau\varepsilon_0}{3}, \quad \forall\ i \neq j,
\end{aligned} \tag{2.3.67}$$

in contradiction with the complete continuity of A.

(ii) Similarly, if (ii) is not true, there exist $h_i \in S_1$ $(i = 1,2,3,\ldots)$ and $\varepsilon_0 > 0$ such that

$$\|A_+'(\infty)h_i - A_+'(\infty)h_j\| > \varepsilon_0 \quad (i = h). \tag{2.3.68}$$

We must have $\sigma = \inf_i \|h_i\| > 0$, since otherwise, there is a subsequence $\{h_{i_k}\} \subset \{h_i\}$ such that $h_{i_k} \to \theta$, and therefore $\|A_+'(\infty)h_{i_k} - A_+'(\infty)h_{i_s}\| \to 0$ as $k, s \to \infty$, in contradiction with (2.3.68). Now, choose $\rho > 0$ such that

$$\|Ax - A_+'(\infty)x\| < \frac{\varepsilon_0}{3}\|x\|, \quad \forall\ x \in P, \quad \|x\| \geq \rho\sigma. \tag{2.3.69}$$

As in (i), it follows from (2.3.68) and (2.3.69) that

$$\|A(\rho_{h_i}) - A(\rho_{h_j})\| \geq \frac{\rho\varepsilon_0}{3}, \quad \forall\ i \neq j,$$

in contradiction with the complete continuity of . □

Corollary 2.3.2. Let $A : P \to P$ be completely continuous and P be generating. Then

(i') if A is differentiable at x_0 P along P, then $A_+'(x_0) : E \to E$ is completely continuous;

(ii') if A is differentiable at infinity along P, then $A_+'(\infty) : E \to E$ is completely continuous.

Proof. This corollary follows directly from Lemma 2.3.5 and Lemma 1.4.2. □

Lemma 2.3.6. Let $A : P \to P$ be completely continuous. Then

(a) if A is differentiable at θ along P, and $A\theta = \theta$ then $A_+'(\theta) : P \to P$ is completely continuous;

(b) if A is differentiable at infinity along P, then $A_+'(\infty) : P \to P$ is completely continuous.

Proof. By Lemma 2.3.5, we need only prove that $A_+'(\theta)$ and $A_+'(\infty)$ map P into P.

Since $A\theta = \theta$, we have for $x > \theta$, $A'_+(\theta)x = \lim_{t \to +0} A(tx)/t$, and hence $A'_+(\theta)x \in P$. Similarly, for $x > \theta$, $A'_+(\infty)x = \lim_{t \to +\infty} A(tx)/t$, and therefore $A'_+(\infty)x \in P$. □

Lemma 2.3.7. Let $A : P \to P$ be completely continuous and $A\theta = \theta$. Suppose that A is differentiable at θ along P and 1 is not an eigenvalue of $A'_+(\theta)$ corresponding to a positive eigenvector. Then

(i) if $A'_+(\theta)$ has no positive eigenvectors corresponding to an eigenvalue greater than one, then there exists $\tau > 0$ such that

$$i(A, P_r, P) = 1, \quad \forall\ 0 < r \leq \tau, \tag{2.3.70}$$

where $P_r = P \cap B_r$ and $B_r = \{x \in E \mid \|x\| < r\}$.

(ii) if $A'_+(\theta)$ has a positive eigenvector corresponding to an eigenvalue greater than one, then there exists $\tau > 0$ such that

$$i(A, P_r, P) = 0, \quad \forall\ 0 < r \leq \tau. \tag{2.3.71}$$

Proof. It is easy to show that

$$\alpha = \inf_{x \in P, \|x\|=1} \|x - A'_+(\tau)x\| > 0. \tag{2.3.72}$$

In fact, if there are $x_n \in P$, $\|x_n\| = 1$ $(n = 1,2,3,\ldots)$ such that $\|x_n - A'_+(\theta)x_n\| \to 0$, then, by Lemma 2.3.6, there exists a subsequence $\{x_{n_k}\}$ such that $A'_+(\theta)x_{n_k} \to x^*$, hence $x_{n_k} \to x^*$, and therefore $x^* \in P$, $\|x^*\| = 1$ and $x^* - A'_+(\theta)x^* = \theta$, in contradiction with the hypotheses.

By virtue of (2.3.72) we know

$$\|x - A'_+(\theta)x\| \geq \alpha\|x\|, \ \forall\ x \in P. \tag{2.3.73}$$

Now, choose $\tau > 0$ such that

$$\|Ax - A'_+(\theta)x\| \leq \frac{\alpha}{2} \|x\|, \quad \forall\ x \in P, \ \|x\| \leq \tau. \tag{2.3.74}$$

It follows from (2.3.73) and (2.3.74) that

$$\|x - [tAx + (1 - t)A'_+(\theta)x]\| \geq \|x - A'_+(\theta)x\| - t\|Ax - A'_+(\theta)x\|$$

$$\geq \alpha\|x\| - \frac{\alpha}{2}\|x\| = \frac{\alpha r}{2} > 0, \quad \forall\ (t,x) \in [0,1]\times(\partial P_r), \quad 0 < r \leq \tau,$$

and hence, by the homotopy invariance property of the fixed point index,

$$i(A,P_r,P) = i(A'_+(\theta),P_r,P), \ \forall\ 0 < r \leq \tau. \tag{2.3.75}$$

In case (i), we know from Lemma 2.3.1 that

$$i(A'(\theta),P_r,P) = 1, \ \forall\ r > 0, \tag{2.3.76}$$

which implies that (2.3.70) follows from (2.3.75) and (2.3.76).

In case (ii), there is $h > \theta$ such that $A'_+(\theta)h = \lambda h$, $\lambda > 1$. We now prove

$$x - A'_+(\theta)x = th, \ \forall\ x > \theta, \quad t \geq 0. \tag{2.3.77}$$

In fact, if there exist $x_0 > \theta$ and $t_0 \geq 0$ such that $x_0 - A'_+(\theta)x_0 = t_0h$, then, by hypotheses, $t_0 > 0$. Since $A'_+(\theta)x_0 \geq \theta$ by Lemma 2.3.6, we have $x_0 \geq t_0h$. Let $t^* = \sup\{t > 0 \mid x_0 \geq th\}$, then $0 < t_0 \leq t^* < +\infty$, $x_0 \geq t^*h$, and so

$$x_0 = A'_+(\theta)x_0 + t_0h \geq A'_+(\theta)(t^*h) + t_0h = (\lambda t^* + t_0)h,$$

which contradicts the definition of t^* since $\lambda t^* + t_0 > t^*$.

By virtue of (2.3.77) and Corollary 2.3.1, we know

$$i(A'_+(\theta),P_r,P) = 0, \ \forall\ r > 0. \tag{2.3.78}$$

consequently, (2.3.71) follows from (2.3.75) and (2.3.78). □

By the same way, we can prove the following.

Lemma 2.3.8. Let $A : P \to P$ be completely continuous. Suppose that A is differentiable at infinity along P and 1 is not an eigenvalue of $A'_+(\infty)$ corresponding to a positive eigenvector. Then

(i') If $A'_+(\infty)$ has no positive eigenvectors corresponding to an eigenvalue greater than one, there exists $\sigma > 0$ such that

$$i(A, P_R, P) = 1, \quad R \geq \sigma, \tag{2.3.79}$$

where $P_R = P \cap B_R$, $B_R = \{x \in E \mid \|x\| < R\}$.

(ii') If $A'_+(\infty)$ has a positive eigenvector corresponding to an eigenvalue greater than one, there exists $\sigma > 0$ such that

$$i(A, P_R, P) = 0, \quad R \geq \sigma. \tag{2.3.80}$$

Theorem 2.3.8. Let $A : P \to P$ be completely continuous and $A\theta = \theta$. Suppose that A is differentiable at θ and at infinity along P, and 1 is not an eigenvalue of $A'_+(\theta)$ or of $A'_+(\infty)$ corresponding to a positive eigenvector. Suppose, moreover, that one of the following two conditions is satisfied:

(H_9) $A'_+(\theta)$ has no positive eigenvectors corresponding to an eigenvalue greater than one, whereas $A'_+(\infty)$ possesses such a positive eigenvector.

(H_{10}) $A'_+(\theta)$ possesses a positive eigenvector corresponding to an eigenvalue greater than one, but it is not the case for $A'_+(\infty)$.

Then A has at least one positive fixed point.

Proof. In the case of (H_9), we know by Lemmas 2.3.7 and 2.3.8 that there exist $R > r > 0$ such that

$$i(A, P_r, P) = 1, \quad i(A, P_R, P) = 0,$$

which implies

$$i(A,P_R\setminus\overline{P}_r,P) = i(A,P_R,P) - i(A,P_r,P) = -1 \neq 0,$$

and therefore, A has at least one fixed point in $P_R\setminus\overline{P}_r$.

Similarly, in the case of (H_{10}), we have

$$i(A,P_r,P) = 0,\ i(A,P_R,P) = 1,\ i(A,P_R\setminus\overline{P}_r,P) = 1 \neq 0 .$$

We can therefore assert that A has at least one fixed point in $P_R\ \overline{P}_r$, too.□

Theorem 2.3.8 is also true for strict-set-contractions (see Cac and Gupta [1]).

Example 2.3.4. Consider the Uryson integral operator (2.3.56). Let $k(t,s,x)$ and $k'_x(t,s,x)$ be continuous and, $k(t,s,x)$ be nonnegative for $t \in G$, $s \in G$, $x \geq 0$, and $k(t,s,0) \equiv 0$. Let (2.3.58) be satisfied uniformly with respect to $t \in G$ and $s \in G$ as $x \to +\infty$. Suppose that 1 is not an eigenvalue of B or of B_1 corresponding to a nonnegative eigenfunction, where

$$Bx(t) = \int_G k'_x(t,s,0)x(s)ds,\ B_1x(t) = \int_G Q(t,s)x(s)ds \ . \tag{2.3.81}$$

Suppose, moreover, that one of the following two conditions is satisfied:

(a) B has no nonnegative eigenfunctions corresponding to an eigenvalue greater than one, whereas B_1 possesses such a nonnegative eigenfunction.

(b) B possess a nonnegative eigenfunction corresponding to an eigenvalue greater than one, but it is not the case for B_1.

Then the Uryson operator (2.3.56) has a fixed point $x(t)$, which is nonnegative, continuous, and not identically zero on G.

The above conclusion follows directly from Theorem 2.3.8 since, by Example 2.3.3, $A'_+(\infty) = B$ and $A'_+(\infty) = B_1$.

2.4 *Multiple Fixed Point Theorems.*

We shall first state the following two theorems on multiple fixed points which are direct consequences of Theorems 2.3.3 and 2.3.4.

Theorem 2.4.1. Let Ω_1, Ω_2 and Ω_3 be three bounded open sets in E such that $\theta \in \Omega_1$, $\overline{\Omega}_1 \subset \Omega_2$, $\overline{\Omega}_2 \subset \Omega_3$. Let operator $A : P \cap (\overline{\Omega}_3\ \Omega_1) \to P$ be completely continuous. Suppose that $Ax \nleq x$, $\forall\ x \in P \cap \partial\Omega_1$; $Ax \ngeq x$, $\forall\, x \in P \cap \partial\Omega_2$ and $Ax \nleq x$, $\forall\ x \in P \cap \partial\Omega_3$. Then A has at least two fixed points x^* and x^{**} in $P \cap (\Omega_3 \setminus \overline{\Omega}_1)$; moreover, $x^* \in \Omega_2 \setminus \overline{\Omega}_1$ and $x^{**} \in \Omega_3 \setminus \overline{\Omega}_2$.

Theorem 2.4.2. Let Ω_1, Ω_2 and Ω_3 be three bounded open sets in E such that $\theta \in \Omega_1$, $\overline{\Omega}_1 \subset \Omega_2$, and $\overline{\Omega}_2 \subset \Omega_3$. Let $A : P \cap (\overline{\Omega}_3 \setminus \Omega_1) \to P$ be completely continuous. Suppose that $\|Ax\| \geq \|x\|$, $\forall\, x \in P \cap \partial\Omega_1$; $\|Ax\| \leq \|x\|$ and $Ax \neq x$, $\forall\, x \in P \cap \partial\Omega_2$; $\|Ax\| \geq \|x\|$, $\forall\, x \in P \cap \partial\Omega_3$. Then A has at least two fixed points x^* and x^{**} in $P \cap (\overline{\Omega}_3 \setminus \Omega_1)$; moreover, $x^* \in \Omega_2\ \Omega_1$ and $x^{**} \in \overline{\Omega}_3 \setminus \overline{\Omega}_2$.

Example 2.4.1. Consider the two-point boundary value problem of an ordinary differential equation:

$$\begin{cases} x'' + x^{\alpha} + x^{\beta} = 0, & 0 \leq t \leq 1; \\ x(0) = x'(1) = 0, & \end{cases} \tag{2.4.1}$$

where $\beta > 1 > \alpha > 0$. Then, the problem (2.4.1) has two nontrivial solutions $x_1(t)$ and $x_2(t)$, which belong to $C^2[0,1]$ and satisfy

$$x_1(t) > 0, \quad x_2(t) > 0, \quad \forall\ 0 < t \leq 1, \tag{2.4.2}$$

$$r < \max_{0 \leq t \leq 1} x_1(t) < 1 < \max_{0 \leq t \leq 1} x_2(t) < R, \tag{2.4.3}$$

where

$$r = \left[\frac{\alpha^{\alpha}}{(\alpha + 2)^{\alpha+2}}\right]^{\frac{1}{2(1-\alpha)}}, \quad R = \left[\frac{(\beta + 2)^{\beta+2}}{\beta^{\beta}}\right]^{\frac{1}{2(\beta-1)}} . \tag{2.4.4}$$

Proof. It is well known that the problem (2.4.1) is equivalent to the integral equation

$$x(t) = \int_0^1 G(t,s)\cdot\{[x(s)]^{\alpha} + [x(s)]^{\beta}\}ds , \tag{2.4.5}$$

where $G(t,s)$ is the Green function of the differential operator $-x''$ with boundary condition $x(0) = x'(1) = 0$, given by

$$G(t,s) = \min\{t,s\} = \begin{cases} t, & t \leq s; \\ s, & t > s. \end{cases} \tag{2.4.6}$$

Let $E = C[0,1]$, $P = \{x(t) \in C[0,1] \mid x(t) \geq 0, \quad 0 \leq t \leq 1\}$ and $P_{\tau} = \{x(t) \in P \mid \min_{\tau \leq t \leq 1} x(t) \geq \tau\|x\|\}$, where $0 < \tau < 1$. Obviously, P and P_{τ} are cones of E and $P_{\tau} \subset P$. Letting

$$Ax(t) = \int_0^1 G(t,s)\cdot\{[x(s)]^{\alpha} + [x(s)]^{\beta}\}ds , \tag{2.4.7}$$

it is clear that $A : P \to P$ is completely continuous. Now, for $x(t) \in P$ and $0 < \tau < 1$, we have

$$\begin{aligned}\min_{\tau \leq t \leq 1} Ax(t) &= \int_0^1 G(\tau,s)\cdot\{[x(s)]^{\alpha} + [x(s)]^{\beta}\}ds \\ &= \int_0^{\tau} s\{[x(s)]^{\alpha} + [x(s)]^{\beta}\}ds + \int_{\tau}^1 \tau\{[x(s)]^{\alpha} + [x(s)]^{\beta}\}ds \\ &\geq \tau \int_0^1 s\{[x(s)]^{\alpha} + [x(s)]^{\beta}\}ds \\ &= \tau \int_0^1 G(1,s)\cdot\{[x(s)]^{\alpha} + [x(s)]^{\beta}\}ds \\ &= \tau\|Ax\| .\end{aligned}$$

Hence $A(P) \subset P_\tau$, and so

$$A(P_\tau) \subset P_\tau, \quad \forall\ 0 < \tau < 1. \tag{2.4.8}$$

In the following, denote the sphere $\{x(t) \in C[0,1] \mid \|x\| = \rho\}$ by S_ρ $(0 < \rho < +\infty)$. For $x(t) \in P \cap S_1$, we have

$$\|Ax\| = \int_0^1 s\{[x(s)]^\alpha + [x(s)]^\beta\}ds \leq (\|x\|^\alpha + \|x\|^\beta)\int_0^1 sds = 1 = \|x\|. \tag{2.4.9}$$

Now we prove

$$Ax \neq x, \quad \forall\ x \in P \cap S_1. \tag{2.4.10}$$

In fact, if there is $x^* \in P \cap S_1$ such that $Ax^* = x^*$, then from (2.4.9),

$$\int_0^1 s\{[x^*(s)]^\alpha + [x^*(s)]^\beta\}ds = 1,$$

and so

$$\int_0^1 s\{(1 - [x^*(s)]^\alpha) + (1 - [x^*(s)]^\beta)\}ds = 0.$$

Consequently, $x^*(s) \equiv 1$ $(0 \leq s \leq 1)$, which contradicts

$$x^*(0) = Ax^*(0) = \int_0^1 G(0,s)\{[x^*(s)]^\alpha + [x^*(s)]^\beta\}ds = 0 .$$

Thus, (2.4.10) holds.

For $x(t) \in P_\tau$, we have

$$\begin{aligned}\|Ax\| &= \int_0^1 s\{[x(s)]^\alpha + [x(s)]^\beta\}ds \geq \int_\tau^1 s\{[x(s)]^\alpha + [x(s)]^\beta\}ds \\ &\geq (\tau^\alpha\|x\|^\alpha + \tau^\beta\|x\|^\beta)\int_\tau^1 sds = \tfrac{1}{2}(1 - \tau^2)(\tau^\alpha\|x\|^\alpha + \tau^\beta\|x\|^\beta).\end{aligned} \tag{2.4.11}$$

It is easy to show that the function $h_2(\tau) = (1 - \tau^2)\tau^\alpha$ attains its maximum in $0 < \tau < 1$ at $\tau = \tau_1 = (\alpha/\alpha + 2)^{1/2}$ and the function

$h_2(\tau) = (1 - \tau^2)\tau^\beta$ at $\tau_2 = (\beta/\beta + 2)^{1/2}$. Putting $\tau = \tau_1$, $\tau = \tau_2$ in (2.4.11), the relations (2.4.4) give

$$x \in P_{\tau_1} \cap S_r \quad \|Ax\| > \frac{1}{2}(1 - \tau_1^2)\tau_1^\alpha \|x\|^\alpha = r = \|x\|, \tag{2.4.12}$$

$$x \in P_{\tau_2} \cap S_R \quad \|Ax\| > \frac{1}{2}(1 - \tau_2^2)\tau_2^\beta \|x\|^\beta = R = \|x\|. \tag{2.4.13}$$

Finally, from (2.4.8), (2.4.9), (2.4.10), (2.4.12), and (2.4.13) we know by Theorem 2.3.4 that A has two fixed points $x_1(t) \in P_{\tau_1}$ and $x_2(t) \in P_{\tau_2}$ such that $r < \|x_1\| < 1 < \|x_2\| < R$.

Lemma 2.4.1. Let X be a retract of the real Banach space E and X_1 be a bounded convex retract of X. Let U be a nonempty open set of X and $U \subset X_1$. Suppose that $A : X_1 \to X$ is completely continuous, $A(X_1) \subset X_1$ and A has no fixed points on $X_1 \backslash U$. Then

$$i(A,U,X) = 1. \tag{2.4.14}$$

Proof. Since the retract of a Hausdorff space must be a closed set, X_1 is closed in X and $\overline{U} \subset X_1$. Thus, by the permanence property of fixed point index, we have

$$i(A,U,X) = i(A,U,X_1). \tag{2.4.15}$$

It is evident, U is an open set of X_1 and hence, by the excision property of fixed point index,

$$i(A,X_1,X_1) = i(A,U,X_1). \tag{2.4.16}$$

Choose $x_0 \in U \subset X_1$ and let

$$H(t,x) = tx_0 + (1 - t)Ax.$$

Observing X_1 is convex, we know that $H : [0,1] \times X_1 \to X_1$ and is completely continuous. Regarding X_1 as a bounded open set of X_1 itself, the boundary of X_1 in X_1, i.e., ∂X_1, is empty, so, by the homotopy invariance property and normality property of fixed point index, we get

$$i(A,X_1,X_1) = i(x_0,X_1,X_1) = 1. \tag{2.4.17}$$

Finally, (2.4.14) follows from (2.4.15), (2.4.16), and (2.4.17). □

Corollary 2.4.1. Let X be a nonempty bounded closed convex set of the real Banach space E. Suppose that $A : X \to X$ is completely continuous. Then $i(A,X,X) = 1$.

Proof. We get this corollary directly by setting $U = X_1 = X$ in Lemma 2.4.1. □

Theorem 2.4.3. Let P be a normal solid cone in the real Banach space E and $y_1, z_1, y_2, z_2 \in E$ with $y_1 < z_1 < y_2 < z_2$. Suppose that $A : [y_1,z_2] \to E$ is completely continuous and strongly increasing and

$$y_1 \leq Ay_1,\ Az_1 < z_1,\ y_2 < Ay_2,\ Az_2 \leq z_2. \tag{2.4.18}$$

Then A has at least three fixed points x_1, x_2, x_3 in $[y_1,z_2]$ such that $y_1 \leq x_1 \ll z_1$, $y_2 \ll x_2 \leq z_2$ and $y_2 \not\leq x_3 \not\leq z_1$.

Proof. Let $X = [y_1,z_2]$, $X_1 = [y_1,z_1]$, and $X_2 = [y_2,z_2]$, then $X_1 \subset X$ and $X_2 \subset X$. Since X, X_1 and X_2 are all nonempty bounded closed convex sets of E, X is a retract of E and X_1, X_2 are retracts of X. By Theorem 2.1.1, A has a maximal fixed point x_1 in $[y_1,z_1]$. From (2.4.18) we know $y_1 \leq x_1 < z_1$, and so $y_1 \leq x_1 = Ax_1 \ll Az_1 < z_1$. Hence x_1 is an interior point of X_1 in X. Let U_1 be the interior of X_1 in X, then $U_1 \neq \phi$ and A has no fixed points in $X \setminus U_1$. Thus, by Lemma 2.4.1, we have

$$i(A,U_1,X) = 1. \tag{2.4.19}$$

In the same way we can prove that A has a minimal fixed point x_2 in $[y_2,z_2]$ with $y_2 \ll x_2 \leq z_2$ and hence x_2 is an interior point of X_2 in X. Similarly, let U_2 be the interior of X_2 in X. Then $U_2 \neq \phi$, A has no fixed points in $X_2 \setminus U_2$ and

$$i(A,U_2,X) = 1. \tag{2.4.20}$$

Now, by the additivity property of fixed point index, we have

$$i(A,X,X) = i(A,U_1,X) + i(A,U_2,X) + i(A,X\setminus\overline{(U_1 \cup U_2)},X), \tag{2.4.21}$$

and by Corollary 2.4.1,

$$i(A,X,X) = 1. \tag{2.4.22}$$

It follows therefore from (2.4.19), (2.4.20), (2.4.21), and (2.4.22) that

$$i(A,X\setminus\overline{(U_1 \cup U_2)},X) = -1,$$

and hence A has a fixed point x_3 in $X\setminus\overline{(U_1 \cup U_2)}$. Observing that A has no fixed points in $X_1\setminus U_1$ and in $X_2\setminus U_2$, we know $x_3 \in X\setminus(X_1 \cup X_2)$. □

Corollary 2.4.2. Under the conditions of Theorem 2.4.3, A has at least three fixed points x_1^*, x_2^*, and x_3^* in $[y_1,z_2]$ such that $x_1^* \ll x_2^* \ll x_3^*$.

Proof. By Theorem 2.1.1, A has in $[y_1,z_2]$ a minimal fixed point x_1^* and a maximal fixed point x_3^*. By virtue of Theorem 2.4.3, $x_1^* \neq x_3^*$ and there exists another fixed point x_2^* of A such that $x_1^* < x_2^* < x_3^*$. Thus, observing that A is strongly increasing, we have $Ax_1^* \ll Ax_2^* \ll Ax_3^*$, i.e., $x_1^* \ll x_2^* \ll x_3^*$. □

In general, we can prove the following conclusion: Let P be a normal solid cone in real Banach space E and y_i, $z_i \in E$ $(i = 1,2,\ldots,m)$ with

$$y_1 < z_1 < y_2 < z_2 < \dots < y_m < z_m .$$

Suppose that $A : [y_1, z_m] \to E$ is completely continuous and strongly increasing and

$$y_1 \leq Ay_1, Az_1 < z_1, y_2 < Ay_2, Az_2 < z_2, \dots, y_m < Ay_m, Az_m \leq z_m. \qquad (2.4.23)$$

Then A has at least $2m - 1$ fixed points $x_1, x_2, \dots, x_{2m-1}$ in $[y_1, z_m]$ such that

$$y_1 \leq x_1 \ll z_1, y_2 \ll x_2 \ll z_2, \dots, y_{m-1} \ll x_{m-1} \ll z_{m-1} ,$$

$$y_m \ll x_m \leq z_m, y_{i+1} \not\leq x_{m+i} \not\leq z_i \quad (i = 1,2,\dots,m-1).$$

We remark that when $m > 2$, condition (2.4.23) is too rigorous and difficult to verify.

<u>Example 2.4.2</u>. Let Ω be a bounded convex domain in R^n whose boundary $\partial\Omega$ belongs to $C^{2+\mu}$ for some $0 < \mu < 1$. Consider a uniformly elliptic differential operator on $\overline{\Omega}$:

$$Lu = - \sum_{i,j=1}^{n} a_{ij}(x)\frac{\partial^2 u}{\partial x_i \partial x_j} + \sum_{i=1}^{n} b_i(x)\frac{\partial u}{\partial x_i} + c(x)u,$$

i.e., there exists a positive constant μ_0 such that

$$\sum_{i,j=1}^{n} a_{ij}(x)\xi_i\xi_j \geq \mu_0|\xi|^2$$

for any $x \in \overline{\Omega}$, $\xi = (\xi_1,\xi_2,\dots,\xi_n) \in R^n$, where $a_{ij}(x) = a_{ij}(x)$, $c(x) \geq 0$, and all functions $a_{ij}(x)$, $b_i(x)$ and $c(x)$ belong to $C^{\mu}(\overline{\Omega})$. Let $\beta = (\beta_1,\beta_2,\dots,\beta_n)$ be a vector field on $\partial\Omega$ of class $C^{1+\mu}$, which satisfies $\beta\cdot n > 0$, where n denotes the outer unit normal on $\partial\Omega$. We consider the boundary operator of the form:

$$Bu = bu + \delta\beta\cdot\text{grad } u = b(x)u + \delta \sum_{i=1}^{n} \beta_i(x)\frac{\partial u}{\partial x_i} , \qquad (2.4.24)$$

where $b(x) \in C^{1+\mu}(\partial\Omega)$. Assume we have one of the following three cases: (a) $\delta = 0$ and $b(x) \equiv 1$ (Dirichlet boundary operator), (b) $\delta = 1$ and $b(x) \equiv 0$ (Neumann boundary operator), (c) $\delta = 1$ and $b(x) > 0$ (oblique derivative boundary operator). Now, consider the boundary value problem:

$$\begin{cases} Lu = f(x,u), & x \in \Omega; \\ Bu = 0, & x \in \partial\Omega, \end{cases} \tag{2.4.25}$$

where $f(x,u) \in C^1(\bar{\Omega} \times R^1)$. A function $v = v(x) \in C^{2+\mu}(\bar{\Omega})$ is called a subsolution for (2.4.25) if

$$\begin{cases} Lv \leq f(x,v), & x \in \Omega; \\ Bv \leq 0, & x \in \partial\Omega. \end{cases}$$

Similarly, a function $w = w(x) \in C^{2+\mu}(\bar{\Omega})$ is called a supersolution for (2.4.25) if

$$\begin{cases} Lw \geq f(x,w), & x \in \Omega; \\ Bw \geq 0, & x \in \partial\Omega. \end{cases}$$

A subsolution (supersolution) is called a strict subsolution (strict supersolution) if it is not a solution. In the following, we denote by $e = e(x) \in C^{2+\mu}(\bar{\Omega})$ the unique solution of the linear boundary value problem:

$$\begin{aligned} Lu &= 1, & x \in \Omega; \\ Bu &= 0, & x \in \partial\Omega. \end{aligned} \tag{2.4.26}$$

Notice, $e(x) > 0$, $\forall\, x \in \Omega$.

Conclusion. If there exist subsolution $v_1(x)$, supersolution $w_2(x)$, strict supersolution $w_1(x)$, strict subsolution $v_2(x)$, for (2.4.25) and constant $\gamma > 0$ such that

$$-\gamma e(x) \leq v_1(x) \leq w_1(x) \leq v_2(x) \leq w_2(x) \leq \gamma e(x),$$
$$v_1(x) \not\equiv w_1(x), \quad w_1(x) \not\equiv v_2(x), \quad v_2(x) \not\equiv w_2(x) \ . \tag{2.4.27}$$

Then, the boundary value problem (2.4.25) has at least three solutions $u_i(x) \in C^{2+\mu}(\overline{\Omega})$ $(i = 1,2,3)$ satisfying

$$v_1(x) \leq u_1(x) < u_2(x) < u_3(x) \leq w_2(x), \quad \forall\, x \in \Omega. \tag{2.4.28}$$

Proof. First we can assume that $f(x,u)$ is strictly increasing with respect to u on $-\gamma M \leq u \leq \gamma M$, where $M = \max\limits_{x\in\overline{\Omega}} e(x)$. Indeed, since $f(x,u) \in C^1(\overline{\Omega} \times R^1)$, there exists $\omega > 0$ such that

$$f(x,u) - f(x,v) > -\omega(u - v), \ \forall\ x \in \overline{\Omega}, \quad -\gamma M \leq v < u \leq \gamma M \ ,$$

and hence the function $f_\omega(x,u) = f(x,u) + \omega u$ is strictly increasing with respect to u on $-\gamma M \leq u \leq \gamma M$. Evidently, problem (2.4.25) is equivalent to the following problem:

$$(L + \omega)u = f_\omega(x,u), \quad x \in \Omega,$$
$$Bu = 0, \quad x \in \partial\Omega, \tag{2.4.29}$$

and $u = u(x)$ is a subsolution (supersolution) for (2.4.25) if and only if $u(x)$ is a subsolution (supersolution) for (2.4.29). Hence, we may replace (2.4.25) by (2.4.29).

By virtue of the classical theory of elliptic partial differential equations, the linear boundary value problem

$$Lu = v, \quad x \in \Omega,$$
$$Bu = 0, \quad x \in \partial\Omega \tag{2.4.30}$$

has exactly one solution $u = Kv \in C^{2+\mu}(\overline{\Omega})$ for every $v \in C^{\mu}(\overline{\Omega})$ and, the bounded linear operator $K : C^{\mu}(\overline{\Omega}) \to C^{2+\mu}(\overline{\Omega})$ can be extended to a completely continuous linear operator (also denoted by K) from $E = C(\overline{\Omega})$

into $E_e = \{u \in C(\bar{\Omega}) \mid \exists\, \lambda > 0,\ -\lambda e(x) \leq u(x) \leq \lambda e(x)\}$ and K is also e-positive (see Lemma 5.3 of Amann [1]). From Theorem 1.5.1 we know that E_e is a Banach space and the imbedding operator $E_e \to E$ is bounded. Obviously, the Nemitskii operator F defined by $Fu = f(x,u(x))$ maps $[v_1,w_2] \subset E$ into E and is bounded, continuous, and strictly increasing. Since v_1, w_1, v_2, $w_2 \in E_e$ by (2.4.27), we know that operator $KF : [v_1,w_2] \subset E_e \to E_e$ is completely continuous and strongly increasing with respect to the solid cone $P_e = E_e \cap P$, where $P = \{u \in E \mid u(x) \geq 0\}$.

Since v_1 is a subsolution for (2.4.25) and $Fv_1 \in C^{\mu}(\bar{\Omega})$,

$$\begin{cases} L(v_1 - KFv_1) \leq 0, & x \in \Omega; \\ B(v_1 - KFv_1) \leq 0, & x \in \partial\Omega. \end{cases}$$

Hence, by the maximum principle (see Amann [2]), $v_1 \leq KFv_1$. By the same way, we get $KFw_1 \leq w_1$, $v_2 \leq KFv_2$ and $KFw_2 \leq w_2$. Observing that w_1 is a strict supersolution and v_2 is a strict subsolution, we have $w_1 \neq KFw_1$, $v_2 \neq KFv_2$, and hence $KFw_1 < w_1$, $v_2 < KFv_2$.

Now, by Corollary 2.4.2, operator KF has at least three fixed points u_1, u_2, and u_3 in $[v_1,w_2]$ such that $v_1 \leq u_1 \ll u_2 \ll u_3 \leq w_2$ and (2.4.28) holds. Since K maps $C(\bar{\Omega})$ into $C^{1+\lambda}(\bar{\Omega})$ for any $0 < \lambda < 1$ (see Amann [1]), we know that $u_i = KFu_i \in C^{1+\lambda}(\bar{\Omega}) \subset C^{\mu}(\bar{\Omega})$ and so $Fu_i \in C^{\mu}(\bar{\Omega})$. Consequently, $u_i = KFu_i \in C^{2+\mu}(\bar{\Omega})$ and u_i $(i = 1,2,3)$ are solutions of (2.4.25). □

In the following, let P be a cone of the real Banach space E and $P_r = \{x \in P \mid \|x\| < r\}$ $(r > 0)$. Then $\partial P_r = \{x \in P \mid \|x\| = r\}$ and $\bar{P}_r = \{x \in P \mid \|x\| \leq r\}$. We shall consider nonnegative continuous and concave functional α on P, i.e., $\alpha : P \to [0,+\infty)$ is continuous and satisfies

$$\alpha(tx + (1-t)y) \geq t\alpha(x) + (1-t)\alpha(y), \quad \forall\, x,y \in P, \quad 0 \leq t \leq 1. \tag{2.4.31}$$

We denote the set $\{x \in P \mid a \leq \alpha(x),\ \|x\| \leq b\}$ $(b > a > 0)$ by $P(\alpha,a,b)$. Obviously, $P(\alpha,a,b)$ is a bounded closed convex set.

Theorem 2.4.4. Let $A : \overline{P}_c \to \overline{P}_c$ be completely continuous and α be a nonnegative continuous concave functional on P with $\alpha(x) \leq \|x\|$ for any $x \in \overline{P}_c$. Suppose that there exist $0 < d < a < b \leq c$ such that

(i) $\{x \in P(\alpha,a,b) \mid \alpha(x) > a\} \neq \phi$ and $\alpha(Ax) > a$ for $x \in P(\alpha,a,b)\}$;

(ii) $\|Ax\| < d$ for $x \in \overline{P}_d$;

(iii) $\alpha(Ax) > a$ for $x \in P(\alpha,a,c)$ with $\|Ax\| > b$.

Then, A has at least three fixed points in $\overline{P}_c$.

Proof. Let $U_1 = \{x \in \overline{P}_c \mid \|x\| < d\}$ and $U_2 = \{x \in P(\alpha,a,c) \mid \alpha(x) > a\}$. By hypotheses, U_1 and U_2 are two nonempty disjoint bounded open convex subsets of $\overline{P}_c$ and $A(\overline{U}_1) \subset U_1$. Hence, by Lemma 2.4.1, we have

$$i(A,U_1,\overline{P}_c) = 1. \tag{2.4.32}$$

Choose $z_0 \in P(\alpha,a,b)$ with $\alpha(z_0) > a$ and let

$$H(t,x) = tz_0 + (1 - t)Ax .$$

Evidently, $H : [0,1] \times \overline{U}_2 \to \overline{P}_c$ is completely continuous. We now prove $H(t,x) \neq x$, $(t,x) \in [0,1] \times \partial U_2$. In fact, if there exists $(t_0,x_0) \in [0,1] \times \partial U_2$ such that $H(t_0,x_0) = x_0$, then $\alpha(x_0) = a$. In case $\|Ax_0\| > b$, we have, by condition (iii), $\alpha(Ax_0) > a$ and hence

$$\alpha(x_0) = \alpha(t_0z_0 + (1 - t_0)Ax_0) \geq t_0\alpha(z_0) + (1 - t_0)\alpha(Ax_0) > a,$$

a contradiction. In the case $\|Ax_0\| \leq b$, we have

$$\|x_0\| = \|t_0z_0 + (1 - t_0)Ax_0\| \leq b,$$

and so $x_0 \in P(\alpha,a,b)$. By condition (i) we again find that $\alpha(Ax_0) > a$ and as before with $\alpha(x_0) > a$, we find the same contradiction.

Thus, by the homotopy invariance property and the normality property of fixed point index, we obtain

$$i(A,U_2,\overline{P}_c) = i(z_0,U_2,\overline{P}_c) = 1. \tag{2.4.33}$$

On the other hand, by Corollary 2.4.1, we have

$$i(A,\overline{P}_c,\overline{P}_c) = 1. \tag{2.4.34}$$

It follows from (2.4.32), (2.4.33), and (2.4.34) that

$$\begin{aligned} i(A,\overline{P}_c\setminus(U_1 \cup U_2),\overline{P}_c) &= i(A,\overline{P}_c,\overline{P}_c) - i(A,U_1,\overline{P}_c) - i(A,U_2,\overline{P}_c) \\ &= -1. \end{aligned} \tag{2.4.35}$$

Finally, by (2.4.32), (2.4.33), and (2.4.35) we know that there exist $x^{(1)} \in U_1$, $x^{(2)} \in U_2$ and $x^{(3)} \in \overline{P}_c\setminus(U_1 \cup U_2)$ such that $Ax^{(i)} = x^{(i)}$ $(i = 1,2,3)$.

<u>Theorem 2.4.5</u>. Let $A : \overline{P}_c \to P$ be completely continuous and α be a nonnegative continuous concave functional on P with $\alpha(x) \le \|x\|$ for any $x \in \overline{P}_c$. Suppose that there exist $0 < d < a < c$ such that

(a) $\{x \in P(\alpha,a,c) \mid \alpha(x) > a\} \neq \phi$ and $\alpha(Ax) > a$ for $x \in P(\alpha,a,c)$;

(b) $\|Ax\| < d$ for $x \in \overline{P}_d$;

(c) $\alpha(Ax) > \frac{a}{c}\|Ax\|$ for any $x \in \overline{P}_c$ with $\|Ax\| > c$.

Then A has at least two fixed points in $\overline{P}_c$.

<u>Proof</u>. Let $U_1 = \{x \in \overline{P}_c \mid \|x\| < d\}$. Then, as before, (2.4.32) holds and hence A has a fixed point x_1 in U_1. Now, we define

$$Bx = \begin{cases} Ax, & \text{when } x \in \overline{P}_c \text{ and } \|Ax\| \le c; \\ \dfrac{cAx}{\|A\|}, & \text{when } x \in \overline{P}_c \text{ and } \|Ax\| > c. \end{cases}$$

It is clear that $B : \overline{P}_c \to \overline{P}_c$ is completely continuous and satisfies all conditions of Theorem 2.4.4 for $b = c$. Hence, by Theorem 2.4.4, B has a fixed point x_3 in $\overline{P}_c \backslash (\overline{U_1 \cup U_2})$, where $U_2 = \{x \in P(\alpha,a,c) \mid \alpha(x) > a\}$. Obviously, $\overline{U_1 \cup U_2} = \overline{U}_1 \cup \overline{U}_2 = \overline{P}_d \cup P(\alpha,a,c)$ and therefore $\alpha(x_3) < a$. If $\|Ax_3\| > c$, then, by condition (c),

$$a > \alpha(x_3) = \alpha(Bx_3) = \alpha\left(\frac{cAx}{\|Ax_3\|}\right) = \alpha\left(\frac{c}{\|Ax_3\|} Ax_3 + \left(1 - \frac{c}{\|Ax_3\|}\right)\theta\right)$$

$$\geq \frac{c}{\|Ax_3\|} \alpha(Ax_3) + \left(1 - \frac{c}{\|Ax_3\|}\right)\alpha(\theta) \geq \frac{c}{\|Ax_3\|} \alpha(Ax_3) > \frac{c}{\|Ax_3\|} \frac{a}{c} \|Ax_3\| = a,$$

a contradiction. Hence $\|Ax_3\| \leq c$, and so $Ax_3 = Bx_3 = x_3$. □

Example 2.4.3. Consider the two-point boundary value problem of the following ordinary differential equation:

$$\begin{cases} x'' + f(x) = 0, & 0 \leq t \leq 1; \\ x(0) = x'(1) = 0. \end{cases} \tag{2.4.36}$$

Conclusion: Let $f(x)$ be nonnegative and continuous on $x \geq 0$ and

$$\overline{\lim_{x \to +\infty}} \frac{f(x)}{x} < 2. \tag{2.4.37}$$

Suppose that there exist $a > d > 0$ such that $f(x) < 2d$ for $0 \leq x \leq d$ and $f(x) \geq 4a$ for $a \leq x \leq 2a$. Then, the problem (2.4.36) has at least three solutions which are nonnegative and belong to $C^2[0,1]$.

Proof. The problem (2.4.36) is equivalent to the integral equation

$$x(t) = \int_0^1 G(t,s)f(x(s))ds, \tag{2.4.38}$$

where $G(t,s)$ is the Green function (2.4.6). Let

$$Ax(t) = \int_0^1 G(t,s)f(x(s))ds, \tag{2.4.39}$$

$E = C[0,1]$ and $P = \{x(t) \in C[0,1] \mid x(t) \geq 0, \; \forall\; 0 \leq t \leq 1\}$. $A : P \to P$ is clearly completely continuous. Let $\alpha(x) = \min\limits_{1/2 \leq t \leq 1} x(t)$. It is also clear that α is a nonnegative continuous concave functional on P with $\alpha(x) \leq \|x\|$ for any $x(t) \in P$. By (2.4.37), there exist $0 < \sigma < 2$ and $\tau > 0$ such that $f(x) \leq \sigma x$ for $x \geq \tau$ and hence

$$0 \leq f(x) \leq \sigma x + \beta, \quad x \geq 0, \tag{2.4.40}$$

where $\beta = \max\limits_{0 \leq x \leq \tau} f(x)$. Now, we choose

$$r > \max\left\{\frac{\beta}{2 - \sigma}, \; 2a\right\} \tag{2.4.41}$$

and prove that all conditions of Theorem 2.4.4 are satisfied for $b = 2a$ and $c = r$. First we prove $A(\overline{P}_r) \subset \overline{P}_r$. Indeed, when $x \in \overline{P}_r$ by (2.4.40) and (2.4.41) we find

$$\|Ax\| = \int_0^1 sf(x(s))ds \leq \int_0^1 s[\sigma x(s) + \beta]ds$$

$$\leq \frac{1}{2}(\sigma\|x\| + \beta) \leq \frac{1}{2}(\sigma r + \beta) < r,$$

and consequently $Ax \in P_r$.

Obviously, $\{x \in P(\alpha,a,2a) \mid \alpha(x) > a\} \neq \phi$ since, for example, $x_0 \in \{x \in P(\alpha,a,2a) \mid \alpha(x) > a\}$, where $x_0(t) \equiv 3a/2$, $0 \leq t \leq 1$. Moreover, for $x \in P(\alpha,a,2a)$, we have

$$\alpha(Ax) = \int_0^1 G\left(\frac{1}{2},s\right)f(x(s))ds > \int_{1/2}^1 G\left(\frac{1}{2},s\right)f(x(s))ds$$

$$\geq \frac{1}{2}\int_{1/2}^1 4a\; ds = a,$$

showing that condition (i) of Theorem 2.4.4 is satisfied. The verification of condition (ii) is simple, since $x \in \overline{P}_d$ implies

$$\|Ax\| = \int_0^1 sf(x(s))ds < \int_0^1 s\cdot 2d\; ds = d.$$

Finally, for $x \in P(\alpha,a,r)$ with $\|Ax\| > 2a$, we have

$$\begin{aligned}\alpha(Ax) &= \int_0^1 G\left(\frac{1}{2},s\right)f(x(s))ds \\ &= \int_0^{1/2} sf(x(s))ds + \frac{1}{2}\int_{1/2}^1 f(x(s))ds \\ &> \frac{1}{2}\left[\int_0^{1/2} sf(x(s))ds + \int_{1/2}^1 sf(x(s))ds\right] \\ &= \frac{1}{2}\int_0^1 sf(x(s))ds = \frac{1}{2}\|Ax\| > a,\end{aligned}$$

and hence condition (iii) is also verified. Thus, by Theorem 2.4.4, A has at least three fixed points in $\overline{P}_r$. □

2.5 *Fixed Points of Domain Expansion and Compression.*

Let us begin with the following lemmas.

Lemma 2.5.1. Let Ω be a bounded open set in real Banach space E, $\theta \in \Omega$ and $A : \overline{\Omega} \to E$ be condensing. Suppose

$$Ax \neq \mu x, \quad \forall\, x \in \partial\Omega, \quad \mu \geq 1. \tag{2.5.1}$$

Then

$$\deg\,(I - A,\Omega,\theta) = 1. \tag{2.5.2}$$

Proof. Let $h(t,x) = x - tAx$. It is easy to see from (2.5.1) that $\theta \notin h(t,\partial\Omega)$ for $0 \leq t \leq 1$, and so by the homotopy invariance and normality of topological degree for condensing fields (see Appendix A.3), we have

$$\deg\,(I - A,\Omega,\theta) = \deg\,(I,\Omega,\theta) = 1. \ \square$$

Corollary 2.5.1. Let Ω be a bounded open set in a real Banach space E, $\theta \in \Omega$ and $A : \overline{\Omega} \to E$ be condensing. Suppose

$$\|Ax\| \leq \|x\|, \quad Ax \neq x, \quad \forall\, x \in \partial\Omega. \tag{2.5.3}$$

Then (2.5.2) holds.

Proof. It is clear that (2.5.3) implies (2.5.1).

Lemma 2.5.2. Let Ω be a bounded open set in infinite dimensional real Banach space E and $A : \overline{\Omega} \to E$ be completely continuous. Suppose that

(i) $\inf_{x \in \partial\Omega} \|Ax\| > 0$;

(ii) $Ax \neq \mu x, \ \forall\, x \in \partial\Omega, \ 0 < \mu \leq 1.$

Then the Leray-Schauder degree

$$\deg\,(I - A, \Omega, \theta) = 0. \tag{2.5.4}$$

Proof. We first prove that

$$\alpha = \inf_{x \in \partial\Omega,\, 0 \leq \mu \leq 1} \|\mu x - Ax\| > 0. \tag{2.5.5}$$

In fact, if (2.5.5) is not true, then there exist $x_n \in \partial\Omega$ and $\mu_n \in [0,1]$ $(n = 1,2,3,\ldots)$ such that $\|\mu_n x_n - Ax_n\| \to 0$ as $n \to \infty$. By the complete continuity of A and the boundedness of μ_n, there exists a subsequence $\{n_k\}$ of $\{n\}$ such that $\mu_{n_k} \to \mu_0 \in [0,1]$ and $Ax_{n_k} \to y_0 \in E$. Obviously $\mu_0 > 0$, since $\mu_0 = 0$ implies $\|Ax_{n_k}\| \to 0$, which contradicts hypothesis (i). Hence $x_{n_k} \to \mu_0^{-1} y = x_0 \in \partial\Omega$, and therefore $\mu_0 x_0 = Ax_0$, $0 < \mu_0 \leq 1$, in contradiction with hypothesis (ii). This shows that (2.5.5) is true.

By the theory of the Leray-Schauder degree (see Appendix, A.3) we can construct a bounded and continuous operator A from $\overline{\Omega}$ into some finite dimensional subspace E_0 of E such that

$$\|Ax - A_1x\| < \alpha, \quad \forall\ x \in \overline{\Omega}.$$

Let the dimension of E_0 be s and $\{y_1, y_2, \ldots, y_s\}$ be a base of E_0. Since E is infinite dimensional, there exists $y_{s+1} \in E \backslash E_0$ such that $\|y_{s+1}\| = 1$. Let E_1 be the subspace spanned by $y_1, y_2, \ldots, y_s, y_{s+1}$. Evidently, we can regard A_1 as a bounded and continuous operator from $\overline{\Omega}$ into E_1 and hence, by the definition of Leray-Schauder degree, we have

$$\deg\,(I - A, \Omega, \theta) = \deg_{E_1}(I - A_1, \Omega_1, \theta), \tag{2.5.6}$$

where $\Omega_1 = \Omega \cap E_1$ and the right side of (2.5.6) denotes the Brouwer degree in E_1. Since, by virtue of (2.5.5),

$$\begin{gathered} \|\mu x - A_1x\| \geq \|\mu x - Ax\| - \|Ax - A_1x\| > \alpha - \alpha = 0, \\ x \in \partial\Omega_1 \subset \partial\Omega, \quad 0 \leq \mu \leq 1, \end{gathered} \tag{2.5.7}$$

it follows from the homotopy invariance of Brouwer degree that

$$\deg_{E_1}(I - A_1, \Omega_1, \theta) = \deg_{E_1}(-A_1, \Omega_1, \theta). \tag{2.5.8}$$

Consider the constant operator $A_0 : A_0x = y_{s+1}$, $\forall\ x \in \overline{\Omega}_1$. Obviously,

$$\deg_{E_1}(A_0, \Omega_1, \theta) = 0. \tag{2.5.9}$$

We now prove that

$$-tA_1x + (1 - t)A_0x = \theta, \quad \forall\ x \in \partial\Omega_1, \quad 0 \leq t \leq 1. \tag{2.5.10}$$

Indeed, if there exists $x_0 \in \partial\Omega_1$ and $t_0 \in [0,1]$ such that

$$-t_0A_1x_0 + (1 - t_0)A_0x_0 = \theta,$$

then, observing $t_0 \neq 1$ by (2.5.7) and $A_1x_0 \in E_0$, we obtain

$$y_{s+1} = \frac{t_0}{1 - t_0} A_1 x_0 \in E_0,$$

in contradiction with $y_{s+1} \in E \backslash E_0$. Hence, by virtue of (2.5.10) we find

$$\deg_{E_1}(-A_1, \Omega_1, \theta) = \deg_{E_1}(A_0, \Omega_1, \theta). \tag{2.5.11}$$

Finally, (2.5.6), (2.5.8), (2.5.9), and (2.5.11) imply (2.5.4) and our proof is complete. □

Corollary 2.5.2. Let Ω be a bounded open set in infinite dimensional real Banach space E, $\theta \notin \partial\Omega$ and $A : \overline{\Omega} \to E$ be completely continuous. Suppose that

$$\|Ax\| \geq \|x\|, \quad Ax \neq x, \quad \forall\, x \in \partial\Omega. \tag{2.5.12}$$

Then (2.5.4) holds.

Proof. It is sufficient to prove that conditions (i) and (ii) of Lemma 2.5.2 are satisfied. From (2.5.12) and $\theta \notin \partial\Omega$, we obtain

$$\inf_{x \in \partial\Omega} \|Ax\| \geq \inf_{x \in \partial\Omega} \|x\| > 0,$$

which shows that (i) is satisfied. If (ii) is not satisfied, then there exist $x_0 \in \partial\Omega$ and $0 < \mu_0 \leq 1$ such that $Ax_0 = \mu_0 x_0$. From (2.5.12) we know $\mu_0 \neq 1$, and therefore $\|Ax_0\| = \mu_0\|x_0\| < \|x_0\|$, in contradiction with condition (2.5.12). □

Notice that Lemma 2.5.2 is not true for strict-set-contraction A. For example, we consider the operator (2.3.25) in space ℓ^2 again and also let $\Omega = \{x = (x_1, x_2, \ldots) \in \ell^2 \mid \|x\| < 1\}$. As before, A is a 1/2-set-contraction from ℓ^2 into ℓ^2 and it is easy to show that

$$Ax \neq \mu x, \quad \forall\, x \in \partial\Omega, \quad \mu \neq 0\ . \tag{2.5.13}$$

From $\|Ax\| = 1/2\ \|x\|$ and (2.5.13) we see that conditions (i) and (ii) of Lemma 2.5.2 are satisfied. But by (2.5.13) and Lemma 2.5.1 we have $\deg(I - A,\Omega,\theta) = 1$.

Notice also that Lemma 2.5.2 is not true for finite-dimensional space E. For example, let $z = re^{i\theta}$ be the polar coordinate of point z in the plane R^2. We define a completely continuous operator A from R^2 into R^2 by

$$Az = z_1, \quad z = re^{i\theta}, \quad z_1 = re^{i(\theta+\frac{\pi}{4})} . \tag{2.5.14}$$

Let $\Omega = \{z = re^{i\theta} \in R^2 \mid r < 1\}$. It is clear that

$$\|Az\| = \|z\|, \quad \forall\ z \in R^2, \tag{2.5.15}$$

and

$$Az \neq \mu z, \quad \forall\ z \neq \theta, \quad \mu \in R^1. \tag{2.5.16}$$

Hence, A satisfies conditions (i) and (ii) of Lemma 2.5.2, but by (2.5.16) and Lemma 2.5.1 we have $\deg(I - A,\Omega,\theta) = 1$.

<u>Lemma 2.5.3</u>. Let Ω be a bounded open set in infinite dimensional real Banach space E and $A : \overline{\Omega} \to E$ be completely continuous. Suppose that

(i') $Ax \neq \mu x$, $\forall\ x \in \partial\Omega$, $0 \leq \mu \leq 1$;

(ii') the set $\{\|Ax\|^{-1}Ax \mid x \in \partial\Omega\}$ is relatively compact. Then (2.5.4) holds.

<u>Proof</u>. Let $A_1x = \alpha(\|Ax\|^{-1})Ax$, where $\alpha = \sup_{x\in\partial\Omega} \|Ax\| > 0$. By hypotheses, $A_1 : \partial\Omega \to E$ is completely continuous. By the extension theorem, A_1 can be extended to a completely continuous operator from $\overline{\Omega}$ into E. We now prove that A_1 satisfies conditions (i) and (ii) of Lemma 2.5.2. In fact,

$$\inf_{x\in\partial\Omega} \|A_1x\| = \alpha > 0.$$

If there exist $x_0 \in \partial\Omega$ and $0 < \mu_0 \leq 1$ such that $A_1 x_0 = \mu_0 x_0$, then $Ax_0 = \lambda_0 x_0$, where $\lambda_0 = \mu_0 \alpha^{-1} \|Ax_0\|$. Evidently, $0 < \lambda_0 \leq \mu_0 \leq 1$, which contradicts hypothesis (i'). Hence, by Lemma 2.5.2, we have

$$\deg(I - A_1, \Omega, \theta) = 0. \tag{2.5.17}$$

Similar to (2.3.28), we can prove

$$(I - (1 - t)A - tA_1)x \neq \theta, \quad \forall\ x \in \partial\Omega, \quad 0 \leq t \leq 1. \tag{2.5.18}$$

Thus, from (2.5.18) and (2.5.17) we get

$$\deg(I - A, \Omega, \theta) = \deg(I - A_1, \Omega, \theta) = 0. \square$$

Lemma 2.5.4. Let Ω be a bounded open set in infinite dimensional real Banach space E and $A : \overline{\Omega} \to E$ be completely continuous. Suppose that there exist $p_1, \ldots, p_n \in E$ and $\tau > 0$ such that every $x \in \partial\Omega$ corresponds to $i = i(x) \in \{1,2,\ldots,n\}$ satisfying

$$\|Ax - p_i\| \geq \|x - p_i\| + \tau.$$

Then (2.5.4) holds.

Proof. Let $D_k = \{x \in \partial\Omega \mid \|Ax - p_k\| \geq \|x - p_k\| + \tau\}$ $(k = 1,2,\ldots,n)$. Obviously, D_k is closed and, by hypotheses,

$$\bigcup_{k=1}^{n} D_k = \partial\Omega. \tag{2.5.19}$$

Since $A(D_k)$ is relatively compact, there exist $y_1^k, y_2^k, \ldots, y_{n_k}^k \in E$ such that every $y \in A(D_k)$ corresponds $y_i^k \in \{y_1^k, y_2^k, \ldots, y_{n_k}^k\}$ satisfying

$$\|y - y_i^k\| < \frac{\tau}{4}.$$

For every y_i^k, we can choose $f_i^k \in E^*$ such that $\|f_i^k\| = 1$ and

$$f_i^k(y_i^k - p_k) = \|y_i^k - p_k\|. \tag{2.5.20}$$

Now, let $x \in \partial\Omega$ be given. On account of (2.5.19), $x \in D_k$ for some k, and hence

$$\|Ax - p_k\| \geq \|x - p_k\| + \tau. \tag{2.5.21}$$

Choose $i \in \{1,2,\ldots,n_k\}$ such that

$$\|Ax - y_i^k\| < \frac{\tau}{4} . \tag{2.5.22}$$

It follows from (2.5.20), (2.5.21), and (2.5.22) that

$$\begin{aligned} f_i^k(x - Ax) &= f_i^k(x - p_k) + f_i^k(p_k - y_i^k) + f_i^k(y_i^k - Ax) \\ &\leq \|x - p_k\| - \|y_i^k - p_k\| + \|y_i^k - Ax\| \\ &\leq \|x - p_k\| - \|p_k - Ax\| + \|Ax - y_i^k\| + \|y_i^k - Ax\| \\ &< -\frac{1}{2}\tau < 0. \end{aligned}$$

Thus, we have proved that every $x \in \partial\Omega$ corresponds f_i^k such that

$$f_i^k(x - Ax) < 0. \tag{2.5.23}$$

Since E is infinite dimensional, there exists a $h \in E$ with $\|h\| = 1$ such that

$$f_i^k(h) = 0, \quad \forall\ 1 \leq k \leq n, \quad 1 \leq i \leq n_k. \tag{2.5.24}$$

It follows from (2.5.23) and (2.5.24) that

$$x - Ax \neq th, \quad \forall\ x \in \partial\Omega, \quad t \geq 0. \tag{2.5.25}$$

Hence (2.5.4) holds (see Appendix, Theorem A.3.3).

Theorem 2.5.1. (Fixed point theorem of domain expansion and compression.) Let Ω_1 and Ω_2 be two bounded open sets in infinite dimensional real Banach space E such that $\theta \in \Omega_1$ and $\overline{\Omega}_1 \subset \Omega_2$. Let operator $A : \overline{\Omega}_2 \backslash \Omega_1 \to E$ be completely continuous. Suppose that one of the two conditions

(H^1) $\|Ax\| \leq \|x\|$ for $x \in \partial\Omega_1$ and $\|Ax\| \geq \|x\|$ for $x \in \partial\Omega_2$

and

(H^2) $\|Ax\| \geq \|x\|$ for $x \in \partial\Omega_1$ and $\|Ax\| \leq \|x\|$ for $x \in \partial\Omega_2$

is satisfied. Then A has at least one fixed point in $\overline{\Omega}_2 \backslash \Omega_1$.

Proof. We need only prove this theorem under condition (H^1), since the proof is similar when (H^2) is satisfied. By the extension theorem, A has a completely continuous extension (also denoted by A) from $\overline{\Omega}_2$ into E. We may assume that A has no fixed points on $\partial\Omega_1$ and $\partial\Omega_2$. By Corollaries 2.5.1 and 2.5.2, we have

$$\deg(I - A,\Omega_1,\theta) = 1 \quad \text{and} \quad \deg(I - A,\Omega_2,\theta) = 0.$$

Consequently,

$$\begin{aligned}\deg(I - A,\Omega_2\backslash\overline{\Omega}_1,\theta) &= \deg(I - A,\Omega_2,\theta) - \deg(I - A,\Omega_1,\theta) \\ &= -1 \neq 0.\end{aligned}$$

Finally, by the solution property of the Leray-Schauder degree (see Appendix, Theorem A.3.3), we know that A has at least one fixed point in $\Omega_2 \backslash \overline{\Omega}_1$ and our theorem is proved. □

We note that Theorem 2.5.1 is not true for finite-dimensional space E. For example, let A be the operator from R^2 into R^2 defined by (2.5.14). Let $\Omega_1 = \{z = re^{i\theta} \mid r < 1\}$ and $\Omega_2 = \{z = re^{i\theta} \mid r < 2\}$. From (2.5.15) we know that conditions (H^1) and (H^2) are satisfied, but, clearly A has no fixed points in $\overline{\Omega}_2 \backslash \Omega_1$.

From Theorem 2.5.1 we can deduce the following result by the same method we employed to show that Theorem 2.3.4 implies Theorem 2.3.7.

Theorem 2.5.2. Let E be an infinite dimensional real Banach space and $A : E \to E$ be completely continuous and $A\theta = \theta$. Suppose that one of the two conditions

$$(H^3) \quad \lim_{\|x\|\to 0} \|Ax\|/\|x\| = 0, \quad \lim_{\|x\|\to+\infty} \|Ax\|/\|x\| = +\infty$$

and

$$(H^4) \quad \lim_{\|x\|\to 0} \|Ax\|/\|x\| = +\infty, \quad \lim_{\|x\|\to+\infty} \|Ax\|/\|x\| = 0$$

is satisfied. Then the following two conclusions hold:

(a) every $\mu \neq 0$ is an eigenvalue of A, i.e., there exists $x_\mu \in E$ with $x_\mu \neq \theta$ such that $Ax_\mu = \mu x_\mu$;

(b) $\lim_{\mu\to\infty} \|x_\mu\| = +\infty$ under (H^3) and $\lim_{\mu\to\infty} \|x_\mu\| = 0$ under (H^4).

Example 2.5.1. Consider the Hammerstein integral equation

$$\phi(x) = \int_G k(x,y)f(y,\phi(y))dy = A\phi(x), \tag{2.5.26}$$

where G is a bounded closed set in R^n and

$$f(x,u) = \sum_{i=1}^{2m} a_i(x)u^i, \tag{2.5.27}$$

m being a positive integer. Obviously, $\phi(x) \equiv 0$ is the trivial solution of equation (2.5.26).

Conclusion: Suppose that (i) the nonnegative continuous kernel $k(x,y)$ satisfies $\int_G k(x,y)dx > 0$ for every $y \in G$; (ii) the bounded measurable function $a_{2m}(x)$ satisfies $\inf_{x\in G} a_{2m}(x) > 0$ and $a_i(x) \in L^{2m/(2m-i)}(G)$ $(i = 1,2,\ldots,2m-1)$. Moreover,

$$\int_G |a_1(x)|^{2m/(2m-1)}dx < (M^{2m} \text{ mes } G)^{-(1/(2m-1))}, \tag{2.5.28}$$

where $M = \max_{x,y\in G} k(x,y)$. Then the integral equation (2.5.26) has at least one continuous nontrivial solution.

Proof. It is well known (see Appendix A.4), that the operator

$$f\phi(x) = f(x,\phi(x))$$

is bounded and continuous from $L^{2m}(G)$ into $L(G)$, and therefore A is completely continuous from $L^{2m}(G)$ into $C(G)$. We have

$$\|A\phi\|_{L^{2m}} \leq M(\text{mes } G)^{1/2m} \sum_{i=1}^{2m} \|a_i\|_{L^{\frac{2m}{2m-1}}} . \|\phi\|^{i}_{L^{2m}}, \quad \forall\ \phi \in L^{m}(G). \tag{2.5.29}$$

Hence, by hypothesis (2.5.28) we can choose a sufficiently small number $r > 0$ such that

$$\|A\phi\|_{L^{2m}} < \|\phi\|_{L^{2m}}, \quad \forall\ \phi \in L^{2m}(G), \quad \|\phi\|_{L^{2m}} = r. \tag{2.5.30}$$

On the other hand, by a well-known inequality (see Hardy, Littlewood, and Polya [1], Theorem 192), we have

$$\left\{\int_G \left[\int_G k(x,y)a_{2m}(y)[\phi(y)]^{2m}dy\right]^{2m}dx\right\}^{1/2m}$$

$$\geq (\text{mes } G)^{(1/2m)-1} \int_G dx \int_G k(x,y)a_{2m}(y)[\phi(y)]^{2m}dy$$

$$= (\text{mes } G)^{(1/2m)-1} \int_G a_{2m}(y)[\phi(y)]^{2m}dy \int_G k(x,y)dx$$

$$\geq \tau\beta(\text{mes } G)^{(1/2m)-1} \|\phi\|^{2m}_{L^{2m}},$$

where $\tau = \inf_{x \in G} a_{2m}(x) > 0$ and $\beta = \min_{y \in G} \int_G k(x,y)dx > 0$. Hence

$$\|A\phi\|_{L^{2m}} \geq \tau\beta(\text{mes } G)^{(1/2m)-1} \|\phi\|^{2m}_{L^{2m}} - M(\text{mes } G)^{1/2m} \sum_{i=1}^{2m-1} \|a_i\|_{L^{\frac{2m}{2m-i}}} \cdot \|\phi\|^{i}_{L^{2m}}, \tag{2.5.31}$$

and so we can choose a sufficiently large number $R > r$ such that

$$\|A\phi\|_{L^{2m}} > \|\phi\|_{L^{2m}}, \quad \forall \ \phi \in L^{2m}(G), \quad \|\phi\|_{L^{2m}} = R. \tag{2.5.32}$$

From (2.5.30) and (2.5.32) it follows by Theorem 2.5.1 that A has a fixed point $\phi \in L^{2m}(G)$ such that $r < \|\phi\|_{L^{2m}} < R$. Since A maps $L^{2m}(G)$ into $C(G)$, we find $\phi \in C(G)$ and our conclusion is proved. □

Notice that under the conditions mentioned above $f(x,u)$ may not be nonnegative.

2.6 *Notes and Comments.*

Theorem 2.1.1 and 2.1.2 have been adapted from Krasnosel'skii [1], Amann [1] and Krasnosel'skii and Zabreiko [1], while Theorem 2.1.3 and 2.1.4 are new and due to Sun and Sun [1]. Theorem 2.1.5 and 2.1.6, which consider fixed points of decreasing operators, have been taken from Guo [3]. Theorem 2.1.7 and 2.1.8, which deal with coupled fixed points, are new and due to Guo and Lakshmikantham [1]. Theorem 2.2.2 to 2.2.5 discuss the fixed points of e-concave and e-convex operators, which can be found in Krasnosel'skii [1]. The concepts of α-concave and $(-\alpha)$-convex operators were originally introduced by Potter [1]. Theorem 2.2.6 to 2.2.8 are due to Guo [6], and improve some results of Potter [1], Wan [1] and Qin [1]. The notion of the fixed point index in Section 2.3 was introduced and discussed by Nussbaum [1] (see also Amann [1], [2].) For a different way of introducing the fixed point index, see Krasnosel'skii and Zabreiko [1]. Since the concept of the fixed point index is based on the Leray-Schauder degree theory, we give some basic definitions and properties of Leray-Schauder degree in the Appendix in order to help readers unfamiliar with this topic. Theorem 2.3.3, the fixed point theorem of cone expansion and compression, is due to Krasnosel'skii [1], while Theorem 2.3.4, the fixed point theorem of cone expansion and compression with norm type, and Theorem 2.5.1, the fixed point theorem of domain expansion and compression, are due to Guo [10], [12]. We do not know if Theorem 2.5.1 is true for condensing mappings. Sun [2] has proved that it is true for condensing mappings when the bounded open sets are balls and the condition $\|Ax\| \geq \|x\|$ is replaced by $\|Ax\| \geq (1 + \delta)\|x\|$ for some $\delta > 0$. Theorem 2.3.7, 2.5.2 and Lemmas 2.3.2, 2.3.3, 2.5.2 are due to Guo [7], [10]. Lemma 2.5.2 was later also proved by Yu and Browder [1]. See Cronin [2] for a result similar to Theorem 2.5.2. Lemmas 2.3.4 and 2.5.3 are taken from Bai [1]. For Lemma 2.5.4, see Huang [1]. Theorem 2.3.8 is known and can be found in Amann [2]. Finally, Theorem 2.4.3 and Corollary 2.4.2 are due to Amann [1], [2], while Theorem 2.4.4 and 2.4.5 were proved by Leggett and Williams [1]. For related results on multiple fixed points see Leatsch [1], Williams and Leggett [1], Cohen [1], Berger [1], and Guo [11]. For recent results on fixed points of weakly inward operators, see Deimling [2] and Deimling and Hu [1].

CHAPTER 3. APPLICATIONS TO NONLINEAR INTEGRAL EQUATIONS

3.0 *Introduction.*

In this chapter, we use the positive fixed point theory established in Chapter 2 together with some special and useful cones to investigate the positive solutions of nonlinear integral equations.

Section 3.1 deals with integral equations whose nonlinearity are of the polynomial type. These are the most important type in practice. We obtain some results about the number of positive solutions in sublinear case, superlinear case, and mixed case, respectively.

In Section 3.2, positive eigenvalues and eigenvectors of nonlinear integral equations are discussed. We also give some applications to nonlinear ordinary and partial differential equations.

Section 3.3 considers some special nonlinear integral equations arising in science, including nonlinear integral equations in neutron transport theory, nuclear physics, and infectious disease modeling. We obtain some results that have physical meaning.

Finally, Section 3.4 discusses the existence of infinitely many solutions of Hammerstein integral equations by global variational method and also gives some applications of ordinary differential equations to two-point boundary value problems.

3.1 *Integral Equations of Polynomial Type.*

In this section we consider the nonlinear integral equation

$$u(x) = \int_G k(x,y)f(y,u(y))dy, \tag{3.1.1}$$

where G is a bounded closed domain in Euclidean space R^n and

$$f(x,u) = \sum_{i=1}^{m} a_i(x)u^{\alpha_i}, \quad \alpha_i > 0, \quad i = 1,2,\ldots,m. \tag{3.1.2}$$

Obviously, $u(x) \equiv 0$ is the trivial solution of Equation (3.1.1). For the sake of convenience, we formulate some conditions for the kernel $k(x,y)$:

(H_1) $k(x,y)$ is nonnegative and continuous on $G \times G$ and there exist a closed set $G_0 \subset G$ with mes $G_0 > 0$ and $0 < \varepsilon_0 < 1$ such that

$$k(x,y) > 0, \quad \forall \ (x,y) \in G_0 \times G_0,$$

$$k(x,y) \geq \varepsilon_0 k(z,y), \quad \forall \ x \in G_0, \quad y \in G, \quad z \in G.$$

(H_2) $k(x,y)$ is nonnegative and continuous on $G \times G$ and there exists $x_0 \in G$ such that $k(x_0,x_0) > 0$.

(H_3) For any closed ball $T \subset \overset{o}{G}$ ($\overset{o}{G}$ denotes the interior of G), there exists $\varepsilon^* = \varepsilon^*(T) > 0$ such that

$$\int_T k(x,y)dy \geq \varepsilon^* \int_G k(x,y)dy, \quad \forall \ x \in G.$$

In the following, the norm in $C(G)$ is denoted by $\|\cdot\|$ and norm in $L(G)$ is denoted by $\| \ \|_L$.

<u>Lemma 3.1.1</u>. Let $u_n(x) \in C(G)$, $u(x) \in C(G)$, $u_n(x) \geq 0$, $u(x) \geq 0$, and $\|u_n - u\| \to 0$ as $n \to \infty$. Then, for any $\alpha > 0$, we have

$$\|u_n^\alpha - u^\alpha\| \to 0 \quad (n \to \infty). \tag{3.1.3}$$

<u>Proof</u>. By a known inequality (see Hardy, Littlewood, and Polya [1])

$$|x^\alpha - y^\alpha| \leq \alpha(x^{\alpha-1} + y^{\alpha-1})|x - y|, \quad x > 0, \quad y > 0, \quad \alpha > 0, \tag{3.1.4}$$

We see that (3.1.3) holds for $\alpha \geq 1$. Now, assume that $0 < \alpha < 1$. For any given $\varepsilon > 0$, let $\varepsilon^* = \left(\frac{\varepsilon}{2\alpha + 2}\right)^{1/\alpha}$, and

$$v(x) = \begin{cases} u(x), & \text{when } u(x) \geq \varepsilon^* \text{ at point } x; \\ \varepsilon^*, & \text{when } u(x) < \varepsilon^* \text{ at point } x, \end{cases}$$

$$w(x) = \begin{cases} u(x), & \text{when } u(x) < \varepsilon^* \text{ at point } x; \\ \varepsilon^*, & \text{when } u(x) \geq \varepsilon^* \text{ at point } x, \end{cases}$$

$$v_n(x) = \begin{cases} u_n(x), & \text{when } u_n(x) \geq \varepsilon^* \text{ at point } x; \\ \varepsilon^*, & \text{when } u_n(x) < \varepsilon^* \text{ at point } x, \end{cases}$$

$$w_n(x) = \begin{cases} u_n(x), & \text{when } u_n(x) < \varepsilon^* \text{ at point } x; \\ \varepsilon^*, & \text{when } u_n(x) \geq \varepsilon^* \text{ at point } x. \end{cases}$$

Obviously,

$$v(x) + w(x) = u(x) + \varepsilon^*, \quad v_n(x) + w_n(x) = u_n(x) + \varepsilon^*,$$

$$[v(x)]^\alpha + [w(x)]^\alpha = [u(x)]^\alpha + (\varepsilon^*)^\alpha,$$

$$[v_n(x)]^\alpha + [w_n(x)]^\alpha = [u_n(x)]^\alpha + (\varepsilon^*)^\alpha,$$

$$\text{and } \|v_n - v\| \to 0, \quad \|w_n - w\| \to 0 \quad (n \to \infty). \tag{3.1.4}$$

Therefore by (3.1.4) we have

$$\begin{aligned} |[u_n(x)]^\alpha - [u(x)]^\alpha| &\leq |[v_n(x)]^\alpha - [v(x)]^\alpha| + |[w_n(x)]^\alpha - [w(x)]^\alpha| \\ &\leq 2\alpha(\varepsilon^*)^{\alpha-1}|v_n(x) - v(x)| + 2(\varepsilon^*)^\alpha. \end{aligned} \tag{3.1.5}$$

From (3.1.4) we know that there exists $N = N(\varepsilon)$ such that

$$\|v_n - v\| < \varepsilon^*, \quad \forall\, n > N. \tag{3.1.6}$$

It follows from (3.1.5) and (3.1.6) that

$$\|u_n^\alpha - u^\alpha\| < 2\alpha(\varepsilon^*)^\alpha + 2(\varepsilon^*)^\alpha = \varepsilon, \quad \forall\, n > N,$$

and hence (3.1.3) holds. □

Theorem 3.1.1. Suppose that (i) the kernel $k(x,y)$ satisfies (H_1), (ii) $a_i(x) \geq 0$, $a_i(x) \in L$ $(n = 1,2,\ldots,m)$ and among α_i $(i = 1,2,\ldots,m)$ there exist $\alpha_{i_0} < 1$ and $\alpha_{i_1} > 1$ such that $\inf_{x \in G_0} a_{i_0}(x) > 0$ and $\inf_{x \in G_0} a_{i_1}(x) > 0$ and (iii) $\sum_{i=1}^{m} \|a_i\|_L < M^{-1}$, where $M = \max_{(x,y) \in G \times G} k(x,y)$. Then, equation (3.1.1) has at least two nontrivial nonnegative continuous solutions.

Proof. Let

$$P = \{u \in C(G) \mid u(x) \geq 0, \ \min_{x \in G_0} u(x) \geq \varepsilon_0 \|u\|\}.$$

It is easy to verify that P is a cone of space $E = C(G)$. We define

$$Au(x) = \int_G k(x,y) f(y,u(y))dy.$$

For $u \in P$, by virtue of (H_1), we have

$$Au(x) \geq \varepsilon_0 \int_G k(z,y) f(y,u(y))dy = \varepsilon_0 Au(z), \quad \forall \ x \in G \ , \quad z \in G,$$

and so

$$\min_{x \in G_0} Au(x) \geq \varepsilon_0 \|Au\| \ ,$$

i.e., $Au \in P$, and therefore

$$A(P) \subset P. \tag{3.1.7}$$

On account of Lemma 3.1.1 it is easy to show that $A : P \to P$ is completely continuous. For $u \in P$ and $x \in G_0$, we have

$$\begin{aligned} Au(x) &\geq \int_{G_0} k(x,y) a_{i_0}(y) [u(y)]^{\alpha_{i_0}} dy \\ &\geq \tau\tau_0 (\text{mes } G) \varepsilon_0^{\alpha_{i_0}} \|u\|^{\alpha_{i_0}} \ , \end{aligned}$$

$$Au(x) \geq \int_{G_0} k(x,y)a_{i_1}(y)[u(y)]^{\alpha_{i_1}}dy$$
$$\geq \tau\tau_1 (\text{mes } G)\varepsilon_0^{\alpha_{i_1}} \|u\|^{\alpha_{i_1}},$$

where $\tau = \min_{(x,y)\in G_0\times G_0} k(x,y) > 0$, $\tau_0 = \inf_{x\in G_0} a_{i_0}(x) > 0$ and $\tau_1 = \inf_{x\in G_0} a_{i_1}(x) > 0$. Hence

$$\|Au\| \geq \tau\tau_0(\text{mes } G)\varepsilon_0^{\alpha_{i_0}}\|u\|^{\alpha_{i_0}}, \quad \forall\ u \in P, \tag{3.1.8}$$

and

$$\|Au\| \geq \tau\tau_1(\text{mes } G)\varepsilon_0^{\alpha_{i_1}}\|u\|^{\alpha_{i_1}}, \quad \forall\ u \in P. \tag{3.1.9}$$

It follows from (3.1.8) and (3.1.9) that there exist $R > 1 > r > 0$ such that

$$\|Au\| > \|u\|, \quad \forall\ u \in P, \quad \|u\| = r \tag{3.1.10}$$

and

$$\|Au\| > \|u\|, \quad \forall\ u \in P, \quad \|u\| = R. \tag{3.1.11}$$

On the other hand, by virtue of condition (iii), we know

$$\|Au\| \leq M \sum_{i=1}^{m} \|a_i\|_L \cdot \|u\|^{\alpha_i} = M \sum_{i=1}^{m} \|a_i\|_L < 1 = \|u\|, \quad \forall\ u \in P, \quad \|u\| = 1. \tag{3.1.12}$$

Now, observing (3.1.7), (3.1.10), (3.1.11), and (3.1.12) and using Theorem 2.3.4, we assert that A has two fixed points u^* and u^{**} in P such that $r < \|u^*\| < 1 < \|u^{**}\| < R$. Obviously, $u^*(x)$ and $u^{**}(x)$ are two nontrivial nonnegative continuous solutions of equation (3.1.1). □

Theorem 3.1.2. Suppose that (i) the kernel $k(x,y)$ satisfies (H_1) and (ii) $a_i(x) \geq 0$, $a_i(x) \in L$, $\alpha_i > 1$ $(i = 1,2,\ldots,m)$ and among $a_i(x)$ $(i = 1,2,\ldots,m)$ there exists $a_{i_0}(x)$ such that $\inf_{x\in G_0} a_{i_0}(x) > 0$. Then, equation (3.1.1) has at least one nontrivial nonnegative continuous solution.

Proof. As in the proof of Theorem 3.1.1 we see that $A : P \to P$ is completely continuous and (3.1.9) holds. Consequently there exists $R > 0$ such that (3.1.11) holds.

On the other hand, since $\alpha_i > 1$ $(i = 1,2,\ldots,m)$, there exists $0 < r < R$ such that

$$\|Au\| \leq M \sum_{i=1}^{m} \|a_i\|_L \cdot \|u\|^{\alpha_i} < \|u\|, \qquad u \in P, \quad \|u\| = r. \tag{3.1.13}$$

It follows from (3.1.11), (3.1.13), and Theorem 2.3.4 that A has at least one fixed point $\bar{u} \in P$ satisfying $r < \|\bar{u}\| < R$. □

Theorem 3.1.3. Suppose that (i') the kernel $k(x,y)$ satisfies (H_2) and (ii") $a_i(x) \geq 0$, $a_i(x) \in L$, $\alpha_i < 1$ $(i = 1,2,\ldots,m)$ and among $a_i(x)$ $(i = 1,2,\ldots,m)$ there exists $a_{i_0}(x)$ such that $\inf_{x \in D_0} a_{i_0}(x) > 0$, where D_0 is a closed ball $D_0 = \{x \in R^n \mid |x - x_0| \leq \delta\} \subset G$ with $k(x_0,x_0) > 0$. Then, there exists $R_0 > 0$ such that, constructing successively the sequence

$$v_j(x) = \int_G k(x,y) f(y, v_{j-1}(y))\,dy \quad (j = 1,2,3,\ldots) \tag{3.1.14}$$

for initial $v_0(x) \equiv R$, where $R \geq R_0$, we have

$$v_0(x) \geq v_1(x) \geq \cdots \geq v_j(x) \geq \cdots, \qquad x \in G \tag{3.1.15}$$

and $v_j(x)$ converges to some $u^*(x)$ as $j \to \infty$ uniformly with respect to $x \in G$ and $u^*(x)$ is a nontrivial nonnegative continuous solution of equation (3.1.1).

Proof. By hypothesis, there exist $\tau > 0$ and $0 < \delta_1 < \delta$ such that

$$k(x,y) \geq \tau, \qquad (x,y) \in D_1 \times D_1, \tag{3.1.16}$$

where $D_1 = \{x \in R^n \mid |x - x_0| \leq \delta_1\}$. Obviously, $D_1 \subset D_0$ and $\tau_1 = \inf_{x \in D_1} a_{i_0}(x) > 0$. Let $D_2 = \{x \in R^n \mid |x - x_0| \leq \delta_1/2\}$ and define a continuous function ψ on G:

$$\psi(x) = \begin{cases} 1, & \text{for } x \in D_2, \\ 0, & \text{for } x \in G \setminus D_1, \\ \text{number between 0 and 1 such that } \psi \text{ is continuous} \\ \text{on } G, & \text{for } x \in D_1 \setminus D_2. \end{cases}$$

Let $\varepsilon' = (\tau\tau_1 \text{ mes } D_2)^{1/(1-\alpha_{i_0})}$ and $0 < \varepsilon \leq \varepsilon'$. As before, define

$$Au(x) = \int_G k(x,y)f(y,u(y))dy \; . \tag{3.1.17}$$

Then, for $x \in D_1$, we have

$$\begin{aligned} A(\varepsilon\psi(x)) &\geq \int_{D_2} k(x,y)a_{i_0}(y)[\varepsilon\psi(y)]^{\alpha_{i_0}}dy \\ &\geq \tau\tau_1\varepsilon^{\alpha_{i_0}} \text{ mes } D_2 \geq \varepsilon \; , \end{aligned}$$

and hence

$$A(\varepsilon\psi(x)) \geq \varepsilon\psi(x), \quad \forall\, x \in G, \quad 0 < \varepsilon \leq \varepsilon'. \tag{3.1.18}$$

On the other hand, we can choose $R_0 > \varepsilon'$ such that

$$M \sum_{i=1}^{m} \|a_i\|_L \cdot R^{\alpha_i - 1} \leq 1, \quad \forall\, R \geq R_0, \tag{3.1.19}$$

where $M = \max\limits_{(x,y)\in G\times G} k(x,y)$. Let $v_0(x) \equiv R$ $(R \geq R_0)$, then, by (3.1.19), we have

$$Av_0(x) \leq M \sum_{i=1}^{m} \|a_i\|_L \cdot R^{\alpha_i} \leq R = v_0(x). \tag{3.1.20}$$

It follows from (3.1.18), (3.1.20), and Theorem 2.1.1 that (3.1.15) holds for sequence (3.1.14) and $v_j(x)$ converges to some $u^*(x)$ as $j \to \infty$ uniformly with respect to $x \in G$ and $Au^*(x) = u^*(x)$, $\varepsilon\psi(x) \leq u^*(x) \leq v_0(x)$. Thus, $u^*(x)$ is a nontrivial nonnegative continuous solution of equation (3.1.1). □

Notice, condition (H_2) is weaker than condition (H_1), i.e., (H_1) implies (H_2). It is clear that the nonnegative continuous kernel $k(x,y)$ satisfies condition (H_1) if there exists $x_0 \in G$ such that $k(x_0,y) > 0$ for any $y \in G$.

Example 3.1.1. Let $k(x,y) = a(x)b(y)$, where $a(x)$ is nonnegative, continuous and nonidentical to zero on G and $b(y)$ is positive and continuous on G. Let $f(x,u) = a_1\sqrt{u} + a_2\sqrt{u^3}$, where $a_1 > 0$ and $a_2 > 0$. Obviously, conditions (i) and (ii) of Theorem 3.1.1 are satisfied. Condition (iii) becomes

$$a_1 + a_2 < (\text{mes } G)^{-1}M_1^{-1}M_2^{-1} , \tag{3.1.21}$$

where $M_1 = \max_{x\in G} a(x)$ and $M_2 = \max_{x\in G} b(x)$. If $u(x)$ is a nontrivial nonnegative continuous solution of equation (3.1.1), then

$$u(x) = a(x) \int_G b(y)\{a_1\sqrt{u(y)} + a_2\sqrt{[u(y)]^3}\}dy , \tag{3.1.22}$$

and hence, $u(x)$ must take the form $u(x) = \gamma a(x)$ with $\gamma > 0$. Substituting $u(x) = \gamma a(x)$ in (3.1.22), we get the equation which must be satisfied by γ:

$$\gamma = a_1b_1\sqrt{\gamma} + a_2b_2\sqrt{\gamma^3} ,$$

or

$$a_2^2b_2^2\gamma^2 + (2a_1a_2b_1b_2 - 1)\gamma + a_1^2b_1^2 = 0, \tag{3.1.23}$$

where

$$b_1 = \int_G b(y)\sqrt{a(y)}dy > 0, \quad b_2 = \int_G b(y)\sqrt{[a(y)]^3}dy > 0.$$

Obviously,

$$b_1 \le (\text{mes } G)M_2\sqrt{M_1}, \quad b_2 \le (\text{mes } G)M_2\sqrt{M_1^3} ,$$

and therefore, by (3.1.21) we find

$$a_1\sqrt{b_1b_2} + a_2\sqrt{b_1b_2} \le (a_1 + a_2)(\text{mes } G)M_1M_2 < 1.$$

From the fact that the function $h(x) = x(1 - x)$ attains maximum $h(1/2) = 1/4$ on $0 < x < 1$ at $x = 1/2$, we see

$$a_1a_2b_1b_2 = a_1\sqrt{b_1b_2}\cdot a_2\sqrt{b_1b_2} < \frac{1}{4} \tag{3.1.24}$$

and so

$$(2a_1a_2b_1b_2 - 1)^2 - 4a_1^2b_1^2a_2^2b_2^2 = 1 - 4a_1a_2b_1b_2 > 0.$$

It follows therefore that the equation (3.1.23) has exactly two different real roots γ_1 and γ_2:

$$\gamma_1 = \frac{1 - 2a_1a_2b_1b_2 - \sqrt{1 - 4a_1a_2b_1b_2}}{2a_2^2b_2^2},$$

$$\gamma_2 = \frac{1 - 2a_1a_2b_1b_2 + \sqrt{1 - 4a_1a_2b_1b_2}}{2a_2^2b_2^2}$$

$(\gamma_2 > \gamma_1 > 0)$, and consequently under condition (3.1.21) equation (3.1.1) has exactly two nontrivial nonnegative continuous solutions $u_1(x) = \gamma_1a(x)$ and $u_2(x) = \gamma_2a(x)$.

Now, we apply Theorem 3.1.1 - 3.1.3 to the following two-point boundary value problems of ordinary differential equations:

$$\begin{cases} -x'' = f(t,x), & 0 \le t \le 1, \\ x(0) = x(1) = 0, & \end{cases} \tag{3.1.25}$$

and

$$\begin{cases} -x'' = f(t,x), & 0 \le t \le 1, \\ x(0) = x'(1) = 0, & \end{cases} \tag{3.1.26}$$

where

$$f(t,x) = \sum_{i=1}^{m} a_i(t)x^{\alpha_i}, \quad \alpha_i > 0, \quad i = 1,2,\ldots,m. \tag{3.1.27}$$

It is well known that problems (3.1.25) and (3.1.26) ($C^2[0,1]$ solution) are equivalent to the integral equations

$$x(t) = \int_0^1 k_1(t,s)f(s,x(s))ds \tag{3.1.28}$$

and

$$x(t) = \int_0^1 k_2(t,s)f(s,x(s))ds \tag{3.1.29}$$

($C^2[0,1]$ solution), respectively, where

$$k_1(t,s) = \begin{cases} t(1-s), & t \leq s; \\ s(1-t), & t > s, \end{cases} \tag{3.1.30}$$

$$k_2(t,s) = \begin{cases} t, & t \leq s; \\ s, & t > s. \end{cases} \tag{3.1.31}$$

We now prove that $k_1(t,s)$ satisfies condition (H_1) for any $G_0 = [\alpha,\beta]$ with $0 < \alpha < \beta < 1$. In fact,

$$k_1(t,s) \geq \alpha(1-\beta), \quad \forall \ (t,s) \in [\alpha,\beta] \times [\alpha,\beta],$$

and for $t \in [\alpha,\beta]$ we have

$$k_1(t,s) = \begin{cases} t(1-s) \geq \alpha(1-s), & \beta \leq s \leq 1; \\ t(1-s) \text{ or } s(1-t) \geq \alpha(1-\beta), & \alpha < s < \beta; \\ s(1-t) \geq (1-\beta)s, & 0 \leq s \leq \alpha. \end{cases}$$

On the other hand,

$$k_1(u,s) \leq s(1-s), \quad \forall \ s \in [0,1], \quad u \in [0,1],$$

and hence

$$k_1(t,s) \geq \alpha(1-\beta)s(1-s) \geq \alpha(1-\beta)k_1(u,s), \quad \forall\, t \in [\alpha,\beta],\ s,\ u \in [0,1],$$

i.e., $k_1(t,s)$ satisfies (H_1) for $G_0 = [\alpha,\beta]$ and $\varepsilon_0 = \alpha(1-\beta)$.

Similarly, we can prove that $k_2(t,s)$ also satisfies (H_1) for any $G_0 = [\alpha,\beta]$ with $0 < \alpha < \beta < 1$ and $\varepsilon_0 = \alpha$.

Thus, by Theorems 3.1.1 - 3.1.3, observing $\max\limits_{0 \leq t,s \leq 1} k_1(t,s) = 1/4$ and $\max\limits_{0 \leq t,s \leq 1} k_2(t,s) = 1$, we obtain the following theorems 3.1.4 and 3.1.5.

<u>Theorem 3.1.4</u>. Let $a_i(t)$ $(i = 1,2,\ldots,m)$ be nonnegative and continuous on $0 \leq t \leq 1$.

(a) if $\alpha_i < 1$ $(i = 1,2,\ldots,m)$ or $\alpha_i > 1$ $(i = 1,2,\ldots,m)$ and $a_i(t)$ $(i = 1,2,\ldots,m)$ are not all identically equal to zero, then problem (3.1.25) has at least one nontrivial nonnegative solution $x(t)$, which belongs to $C^2[0,1]$ and $x(t) > 0$ for $0 < t < 1$;

(b) if $\sum\limits_{i=1}^{m} \int_0^1 a_i(t)dt < 4$ and among α_i $(i = 1,2,\ldots,m)$ there exist $\alpha_{i_0} < 1$ and $\alpha_{i_1} > 1$ such that $a_{i_0}(t)$ and $a_{i_1}(t)$ are not identical to zero, then problem (3.1.25) has at least two nontrivial nonnegative solutions $x_1(t)$ and $x_2(t)$, which belong to $C^2[0,1]$ and $x_1(t) > 0$, $x_2(t) > 0$ for $0 < t < 1$.

<u>Theorem 3.1.5</u>. Let $a_i(t)$ $(i = 1,2,\ldots,m)$ be nonnegative and continuous on $0 \leq t \leq 1$.

(a) if $\alpha_i < 1$ $(i = 1,2,\ldots,m)$ or $\alpha_i > 1$ $(i = 1,2,\ldots,m)$ and $a_i(t)$ $(i = 1,2,\ldots,m)$ are not all identical to zero, then problem (3.1.26) has at least one nontrivial nonnegative solution $x(t)$, which belongs to $C^2[0,1]$ and $x(t) > 0$ for $0 < t \leq 1$;

(b) if $\sum\limits_{i=1}^{m} \int_0^1 a_i(t)dt < 1$ and among α_i $(i = 1,2,\ldots,m)$ there exist $\alpha_{i_0} < 1$ and $\alpha_{i_1} > 1$ such that $a_{i_0}(t)$ and $a_{i_1}(t)$ are not identical to zero, then problem (3.1.26) has at least two nontrivial nonnegative solutions $x_1(t)$ and $x_2(t)$, which belong to $C^2[0,1]$ and $x_1(t) > 0$, $x_2(t) > 0$ for $0 < t \leq 1$.

Theorem 3.1.6. Suppose that (i) the kernel $k(x,y)$ satisfies (H_2) and (H_3) and (ii) nonnegative and continuous functions $a_i(x)$ satisfy $\sum_{i=1}^{m} a_i(x) > 0$ for any $x \in G$ and $\alpha_i < 1$ $(i = 1,2,\ldots,m)$. Then equation (3.1.1) has exactly one nontrivial nonnegative continuous solution $u^*(x)$. Moreover, constructing successively the sequence of functions

$$u_n(x) = \int_G k(x,y)f(y,u_{n-1}(y))dy \quad (n = 1,2,3,\ldots) \tag{3.1.32}$$

for any initial function $u_0(x)$, which is continuous and nonnegative and not identical to zero on G, we have

$$\|u_n - u^*\| \to 0 \quad (n \to \infty). \tag{3.1.33}$$

Proof. It is clear that conditions (i') and (ii") of Theorem 3.1.3 are satisfied, and so Theorem 3.1.3 implies that equation (3.1.1) has at least one nontrivial nonnegative continuous solution $u^*(x)$. Now, suppose that equation (3.1.1) has another such solution $u^{**}(x)$. Then, there exist a closed ball $T \subset \overset{o}{G}$ and $0 < \tau < 1$ such that $u^{**}(x) \geq \tau$ for any $x \in T$. Thus, by condition (H_3), we have

$$u^{**}(x) \geq \int_T k(x,y) \sum_{i=1}^{m} a_i(y)\tau^{\alpha_i} dy \geq \sigma\tau^{\alpha} \int_T k(x,y)dy$$

$$\geq \sigma\tau^{\alpha}\varepsilon^* \int_G k(x,y)dy \ , \quad \forall\ x \in G, \tag{3.1.34}$$

where $\sigma = \min_{x\in G} \sum_{i=1}^{m} a_i(x) > 0$, $\alpha = \max_i \alpha_i < 1$ and ε^* is the positive number defined by (H_3).

On the other hand, letting $\beta_i = \max_{x\in G} a_i(x)$ $(i = 1,2,\ldots,m)$, we find

$$u^*(x) \leq \left(\sum_{i=1}^{m} \beta_i \|u^*\|^{\alpha_i} \right) \int_G k(x,y)dy, \quad \forall\ x \in G. \tag{3.1.35}$$

It follows from (3.1.34) and (3.1.35) that

$$u^{**}(x) \geq \gamma u^*(x), \quad \forall\ x \in G, \tag{3.1.36}$$

where

$$\gamma = \sigma\tau^{\alpha}\varepsilon^{*}\left(\sum_{i=1}^{m} \beta_i \|u^*\|^{\alpha_i}\right)^{-1} > 0 .$$

Now, let $t_0 = \sup\{t > 0 \mid u^{**}(x) \geq tu^*(x)\}$. It is clear that $0 < \gamma \leq t_0 < +\infty$ and $u^{**}(x) \geq t_0 u^*(x)$. We prove $t_0 \geq 1$. In fact, if $0 < t_0 < 1$, then we have

$$\begin{aligned} u^{**}(x) &\geq \int_G k(x,y)\left\{\sum_{i=1}^{m} a_i(y)[t_0 u^*(y)]^{\alpha_i}\right\}dy \\ &\geq t_0^{\alpha} \int_G k(x,y)\left\{\sum_{i=1}^{m} a_i(y)[u^*(y)]^{\alpha_i}\right\}dy \\ &= t_0^{\alpha}\, u^*(x), \quad \forall\, x \in G, \end{aligned}$$

which contradicts the definition of t_0, since $\alpha = \max_i \alpha_i < 1$ and $t_0^{\alpha} > t_0$. Thus, $t_0 \geq 1$ and $u^{**}(x) \geq u^*(x)$ for any $x \in G$.

In the same way, we can prove $u^*(x) \geq u^{**}(x)$ for any $x \in G$, and hence $u^{**}(x) \equiv u^*(x)$ and the uniqueness of nontrivial solutions of equation (3.1.1) is proved.

Finally, we need to prove (3.1.33). For a given nontrivial initial function $u_0(x)$, there exists a closed ball $T_1 \subset \overset{o}{G}$ and $0 < \tau_1 < 1$ such that $u_0(x) \geq \tau_1$ for any $x \in T_1$. Similar to (3.1.34), we can get

$$u_1(x) = Au_0(x) \geq \sigma\tau_1^{\alpha}\varepsilon_1^{*} \int_G k(x,y)dy, \quad \forall\, x \in G, \tag{3.1.37}$$

where A is the operator (3.1.17) and ε_1^* is the positive number defined by (H_3) with respect to T_1. Now, we choose a sufficiently small positive number ε such that

$$\sum_{i=1}^{m} \beta_i \varepsilon^{\alpha_i} \leq \sigma\tau_1^{\alpha}\varepsilon_1^{*} \tag{3.1.38}$$

and (3.1.18) holds. Letting $w_0(x) \equiv \varepsilon\psi(x)$ and $w_n(x) = Aw_{n-1}(x)$ $(n = 1,2,3,\ldots)$, we see from (3.1.18), (3.1.37), and (3.1.38) that

$$w_1(x) = A\,\varepsilon\psi(x) \le \left(\sum_{i=1}^{m} \beta_i \varepsilon^{\alpha_i}\right) \int_G k(x,y)dy$$

$$\le \sigma\tau_1^{\alpha}\varepsilon_1^{*} \int_G k(x,y)dy \le u_1(x), \quad \forall\, x \in G. \tag{3.1.39}$$

On the other hand, choose a number $R \ge \max\{R_0, \max_{x\in G} u_0(x)\}$, where R_0 is defined in Theorem 3.1.3, then $v_0(x) \equiv R \ge u_0(x)$, and so

$$v_1(x) = A[v_0(x)] \ge A[u_0(x)] = u_1(x). \tag{3.1.40}$$

It follows from (3.1.39) and (3.1.40) that

$$w_n(x) \le u_n(x) \le v_n(x) \quad (n = 1,2,3,\ldots), \tag{3.1.41}$$

where $\{v_n(x)\}$ is the sequence (3.1.14). Observing (3.1.18), (3.1.20), and using Theorem 2.1.1, and noticing the uniqueness of nontrivial solutions of equation (3.1.1), we get

$$\|w_n - u^*\| \to 0 \quad \text{and} \quad \|v_n - u^*\| \to 0 \quad \text{as} \quad n \to \infty,$$

which together with (3.1.41) implies (3.1.33).

It is clear that the Green functions (3.1.30) and (3.1.31) satisfy condition (H_2). Now, we prove that $k_1(t,s)$ and $k_2(t,s)$ also satisfy (H_3). In fact,

$$\int_0^1 k_1(t,s)ds = \int_0^t s(1-t)ds + \int_t^1 t(1-s)ds = \frac{1}{2}\,t(1-t).$$

For any $0 < \xi < \eta < 1$, we have

$$\int_\xi^\eta k_1(t,s)ds \begin{cases} = \displaystyle\int_\xi^\eta s(1-t)ds \ge \xi(\eta-\xi)(1-x) & \text{for } \eta \le t \le 1; \\ \ge \displaystyle\int_\xi^\eta \xi(1-\eta)ds = \xi(1-\eta)(\eta-\xi) & \text{for } \xi < t < \eta; \\ = \displaystyle\int_\xi^\eta t(1-s)ds \ge (1-\eta)(\eta-\xi)x & \text{for } 0 \le t \le \xi. \end{cases}$$

Hence

$$\int_\xi^\eta k_1(t,s)ds \geq 2\xi(1-\eta)(\eta-\xi)\int_0^1 k_1(t,s)ds, \quad \forall\ 0 \leq t \leq 1,$$

which shows that $k_1(t,s)$ satisfies (H_3) by taking $\varepsilon^* = 2\xi(1-\eta)(\eta-\xi)$ for $T = [\xi,\eta] \subset (0,1) = \overset{o}{G}$.

Similarly, we can get

$$\int_\xi^\eta k_2(t,s)ds \geq \xi(\eta-\xi)\int_0^1 k_2(t,s)ds, \quad \forall\ 0 \leq t \leq 1, \quad 0 < \xi < \eta < 1,$$

and so $k_2(t,s)$ also satisfies (H_3).

Thus, by Theorem 3.1.6, we get the following.

Theorem 3.1.7. Let $\alpha_i < 1$ $(i = 1,2,\ldots,m)$ and $a_i(t)$ $(i = 1,2,\ldots,m)$ be nonnegative and continuous such that $\sum_{i=1}^m a_i(t) > 0$ for any $0 \leq t \leq 1$. Then the following two conclusions hold:

(a) problem (3.1.25) has one and only one nontrivial nonnegative solution $x^*(t)$, which belongs to $C^2[0,1]$ and $x^*(t) > 0$ for $0 < t < 1$; moreover, constructing successively the sequence of functions

$$x_n(t) = (1-t)\int_0^t s\left\{\sum_{i=1}^m a_i(s)[x_{n-1}(s)]^{\alpha_i}\right\}ds$$
$$+ t\int_t^1 (1-s)\left\{\sum_{i=1}^m a_i(s)[x_{n-1}(s)]^{\alpha_i}\right\}ds \quad (n = 1,2,3,\ldots)$$

for any initial function $x_0(t)$, which is continuous and nonnegative and not identical to zero on $[0,1]$, sequence $\{x_n(t)\}$ must converge to $x^*(t)$ uniformly on $0 \leq t \leq 1$;

(b) problem (3.1.26) has one and only one nontrivial nonnegative solution $x^*(t)$, which belongs to $C^2[0,1]$ and $x^*(t) > 0$ for $0 < t \leq 1$; moreover, when we successfully construct the sequence of functions

$$x_n(t) = \int_0^t s\left\{ \sum_{i=1}^{m} a_i(s)[x_{n-1}(s)]^{\alpha_i} \right\} ds$$

$$+ t \int_t^1 \sum_{i=1}^{m} a_i(s)[x_{n-1}(s)]^{\alpha_i} \Bigg\} ds \qquad (n = 1,2,3,\ldots)$$

for any initial function $x_0(t)$, which is continuous, nonnegative and not identical to zero on $[0,1]$, the sequence $\{x_n(t)\}$ must converge to $x^*(t)$ uniformly on $0 \leq t \leq 1$.

Theorem 3.1.8. Suppose that (i) the kernel $k(x,y)$ is continuous and nonnegative on $G \times G$ and satisfies (H_3) and (ii) nonnegative and continuous functions $a_i(x)$ satisfy $\sum_{i=1}^{m} a_i(x) > 0$ for any $x \in G$ and $\alpha_i > 1$ $(i = 1,2,\ldots,m)$. Then equation (3.1.1) cannot have two comparable nontrivial nonnegative continuous solutions.

Proof. Suppose that $u^*(x)$ and $u^{**}(x)$ are two nontrivial nonnegative continuous solutions of equation (3.1.1) such that $u^{**}(x) \geq u^*(x)$ and $u^{**}(x) \equiv u^*(x)$ for $x \in G$. Then, there exists a closed ball $T \subset \overset{o}{G}$ and $0 < \tau < 1$ such that

$$[u^{**}(x)]^{\alpha_i} - [u^*(x)]^{\alpha_i} \geq \tau, \quad \forall\ x \in T, \quad i = 1,2,\ldots,m.$$

Thus, by condition (H_3), we find

$$u^{**}(x) - u^*(x) \geq \int_T k(x,y)\cdot\left\{ \sum_{i=1}^{m} a_i(y)\ (u^{**}(x))^{\alpha_i} - (u^*(y))^{\alpha_i} \right\} dy$$

$$\geq \tau\sigma \int_T k(x,y)dy \geq \tau\sigma\varepsilon^* \int_G k(x,y)dy, \qquad (3.1.42)$$

where $\sigma = \min_{x \in G} \sum_{i=1}^{m} a_i(x) > 0$ and ε^* is the positive number defined by (H_3). On the other hand, (3.1.35) still holds in this case. It follows therefore from (3.1.42) and (3.1.35) that

$$u^{**}(s) \geq (1+\gamma)u^*(x), \quad \forall\ x \in G, \tag{3.1.43}$$

where

$$\gamma = \tau\sigma\varepsilon^*\left(\sum_{i=1}^{m} \beta_i \|u''\|^{\alpha_i}\right)^{-1} > 0.$$

Now, letting $t_0 = \sup\{t > 0 \mid u^{**}(x) \geq tu^*(x)\}$, from (3.1.43) we find $1 < 1 + \gamma \leq t_0 < +\infty$ and $u^{**}(x) \geq t_0 u^*(x)$. As a result

$$u^{**}(x) \geq \int_G k(x,y)\left\{\sum_{i=1}^{m} a_i(y)\ t_0 u^*(y)^{\alpha_i}\right\}dy$$

$$\geq t_0^{\alpha} \int_G k(x,y)\left\{\sum_{i=1}^{m} a_i(y)\ u^*(y)^{\alpha_i}\right\}dy = t_0^{\alpha}u^*(x), \tag{3.1.44}$$

where $\alpha = \min\{\alpha_1,\ldots,\alpha_m\} > 1$. Since $t_0^{\alpha} > t_0$, (3.1.44) contradicts the definition of t_0. □

As a consequence of Theorem 3.1.8, we obtain the following.

Theorem 3.1.9. Suppose that $\alpha_i > 1$ $(i = 1,2,\ldots,m)$ and $a_i(t)$ $(i = 1,2,\ldots,m)$ are nonnegative and continuous such that $\sum_{i=1}^{m} a_i(t) > 0$ for any $0 \leq t \leq 1$. Then either problem (3.1.25) or problem (3.1.26) cannot have two comparable nontrivial solutions.

The following two theorems are concerned with the eigenvalues and eigenfunctions of operator (3.1.17).

Theorem 3.1.10. Suppose that (a) the kernel $k(x,y)$ satisfies (H_1) and (b) $a_i(x) \geq 0$, $a_i(x) \in L$, $\alpha_i > 1$ $(i = 1,2,\ldots,m)$ and among $a_i(x)$ $(i = 1,2,\ldots,m)$ there exists $a_{i_1}(x)$ such that $\inf_{x\in G_0} a_{i_1}(x) > 0$. Then, every $\mu > 0$ is an eigenvalue of operator (3.1.17), which corresponds to nonnegative eigenfunction, i.e., there exists nontrivial nonnegative continuous function $u_\mu(x)$ such that

$$\int_G k(x,y)f(y,u_\mu(y))dy = \mu u_\mu(x).$$

Moreover, $\max_{x \in G} u_\mu(x) \to +\infty$ as $\mu \to +\infty$.

Theorem 3.1.11. Suppose that (a) the kernel $k(x,y)$ satisfies (H_1) and (b') $a_i(x) \geq 0$, $a_i(x) \in L$, $\alpha_i < 1$ $(i = 1,2,\ldots,m)$ and among $a_i(x)$ $(i = 1,2,\ldots,m)$ there exists $a_{i_0}(x)$ such that $\inf_{x \in G_0} a_{i_0}(x) > 0$. Then, every $\mu > 0$ is an eigenvalue of operator (3.1.17), which corresponds to nonnegative eigenfunction, i.e., there exists nontrivial nonnegative continuous function $u_\mu(x)$ such that (3.1.45) holds. Moreover, $\max_{x \in G} u_\mu(x) \to 0$ as $\mu \to +\infty$.

Proof of Theorems 3.1.10 and 3.1.11. As in the proof of Theorem 3.1.1 we see that $A : P \to P$ is completely continuous, where

$$P = \{u \in C(G) \mid u(x) \geq 0, \ \min_{x \in G_0} u(x) \geq \varepsilon_0 \|u\|\},$$

and (3.1.9) holds under the conditions of Theorem 3.1.10 and (3.1.8) holds under the conditions of Theorem 3.1.11. Hence

$$\lim_{u \in P, \|u\| \to +\infty} \frac{\|Au\|}{\|u\|} = +\infty \tag{3.1.46}$$

and

$$\lim_{u \in P, \|u\| \to 0} \frac{\|Au\|}{\|u\|} = +\infty \tag{3.1.47}$$

hold under the conditions of Theorem 3.1.10 and Theorem 3.1.11, respectively. On the other hand, we have

$$\|Au\| \leq M \sum_{i=1}^{m} \|a_i\|_L \cdot \|u\|^{\alpha_i}, \quad \forall\, u \in P, \tag{3.1.48}$$

where

$$M = \max_{x,y \in G} k(x,y) ,$$

and therefore

$$\lim_{u\in P, \|u\|\to 0} \frac{\|Au\|}{\|u\|} = 0 \tag{3.1.49}$$

under conditions of Theorem 3.1.10, and

$$\lim_{u\in P, \|u\|\to +\infty} \frac{\|Au\|}{\|u\|} = 0 \tag{3.1.50}$$

under conditions of Theorem 3.1.11. Hence, observing (3.1.49), (3.1.46), and (3.1.47), (3.1.50), the conclusions of Theorems 3.1.10 and 3.1.11 follow from Theorem 2.3.7. □

3.2 *Eigenvalues and Eigenfunctions.*

This section deals with the nontrivial solutions of the nonlinear integral equations

$$\lambda u(x) = \int_G k(x,y) f[u(y)]dy = Au(x), \tag{3.2.1}$$

where G is a bounded closed domain in Euclidean space R^n and $f(0) = 0$. Notice that a nontrivial solution $u(x)$ of equation (3.2.1) is called an eigenfunction of operator A corresponding to the eigenvalue λ.

Besides conditions (H_1), (H_2), and (H_3) of section 3.1, we formulate two more conditions for the kernel $k(x,y)$:

(H_4) the nonnegative and continuous kernel $k(x,y)$ is symmetric on $G \times G$ and there exists $(x_0,y_0) \in G \times G$ such that $k(x_0,y_0) > 0$;

(H_5) $k(x,y)$ is symmetric and continuous for $(x,y) \in G \times G$ with $x \neq y$ and satisfies

$$k(x,y) \not\equiv 0, \quad 0 \leq k(x,y) \leq \frac{M}{|x - y|^{\nu}}, \quad x,\ y \in G, \quad x \neq y,$$

where M_0 = const. and $0 < \nu < n$.

In the following, $f'_+(0)$ always denotes the right derivative of function $f(u)$ at $u = 0$.

Theorem 3.2.1. Suppose that (i) the kernel $k(x,y)$ satisfies (H_2) or (H_4) or (H_5) and (ii) the function $f(u)$ is continuous on $0 \leq u \leq \delta$ $(\delta > 0)$, $f(0) = 0$ and $f'_+(0)$ is finite and not equal to zero. Then there exist $\varepsilon_0 > 0$ and $\eta_0 > 0$ such that equation (3.2.1) has a nontrivial and nonnegative continuous solution $u_r(x)$ with $\|u_r\| = r$ and $|\lambda_r| \geq \eta_0$ for any $0 < r < \varepsilon_0$, i.e.,

$$\lambda_r u_r(x) = Au_r(x),\ u_r(x) \in C(G),\ u_r(x) \geq 0,\ \|u_r\| = r,\ |\lambda_r| \geq \eta_0. \tag{3.2.2}$$

Proof. We first assume that $k(x,y)$ satisfies (H_2). If $0 < f'_+(0) < +\infty$, then there exist $0 < \varepsilon_0 < \delta$ and $\sigma > 0$ such that

$$f(u) \geq \sigma u,\quad \forall\ 0 \leq u \leq \varepsilon_0. \tag{3.2.3}$$

By (H_2) we can choose $x_0 \in \overset{o}{G}$ such that $k(x_0,x_0) > 0$ and therefore there exist a closed ball $D_1 = \{x \in R^n \mid |x - x_0| \leq \delta_1\} \subset G$ and $\tau > 0$ such that

$$k(x,y) \geq \tau,\quad \forall\ (x,y) \in D_1 \times D_1. \tag{3.2.4}$$

Let $\eta_0 = \tau\sigma \text{ mes } D_2$, where $D_2 = \{x \in R^n \mid |x - x_0| \leq \delta_1/2\}$. We now prove that ε_0 and η_0 are the numbers to be required. Let $P = \{u \in C(G) \mid u(x) \geq 0\}$ and $B_\ell = \{u \in C(G) \mid \|u\| < \ell\}$, $S_\ell = \{u \in C(G) \mid \|u\| = \ell\}$ for $0 < \ell < +\infty$. Then, P is a solid normal cone in $E = C(G)$ and, by virtue of (3.2.3), $A : P \cap \overline{B}_{\varepsilon_0} \to P$ is completely continuous. Now let $0 < r < \varepsilon_0$ be given. We may assume that

$$u - \eta_0^{-1} Au \neq \theta,\quad \forall\ u \in P \cap S_r, \tag{3.2.5}$$

for otherwise there exists $u_r \in P \cap S_r$ such that $\eta_0 u_r = Au_r$ and therefore (3.2.2) already holds for $\lambda_r = \eta_0$.

Define a continuous function ψ on G as follows:

$$\psi(x) = \begin{cases} 1, & \text{for } x \in D_2; \\ 0, & \text{for } x \in G \setminus D_1; \\ \text{number between } 0 \text{ and } 1 \text{ such that } \psi \text{ is} & \\ \quad \text{continuous on } G, \text{ for } x \in D_1 \setminus D_2. & \end{cases}$$

We now prove

$$u - \eta_0^{-1} Au = t\psi, \ \forall \ u \in P \cap S_r, \quad t \geq 0. \tag{3.2.6}$$

In fact, if there exist $u_0 \in P \cap S_r$ and $t_0 \geq 0$ such that $u_0 - \eta_0^{-1} Au_0 = t_0\psi$, then, by (3.2.5), $t_0 > 0$ and $u_0 \geq t_0\psi$. Letting $t^* = \sup\{t > 0 \mid u_0 \geq t\psi\}$, we see $0 < t_0 \leq t^* < +\infty$ and $u_0 \geq t^*\psi$. Now, for $x \in D_1$, by virtue of (3.2.3) and (3.2.4), we find

$$\begin{aligned} u_0(x) &= \eta_0^{-1} \int_G k(x,y) f\, u_0(y)\, dy + t_0\psi(x) \\ &\geq \eta_0^{-1} \int_{D_2} k(x,y)\sigma t^*\psi(y) dy + t_0\psi(x) \\ &\geq \eta_0^{-1} \tau\sigma t^* \text{mes } D_2 + t_0\psi(x) = t^* + t_0\psi(x), \end{aligned}$$

and hence

$$u_0(x) \geq (t^* + t_0)\psi(x), \quad \forall x \in G,$$

which contradicts the definition of t^*. Thus, (3.2.6) holds and, by corollary 2.3.1, the fixed point index

$$i(\eta_0^{-1} A, P \cap B_r, P) = 0. \tag{3.2.7}$$

On the other hand, it is clear that

$$i(\theta, P \cap B_r, P) = 1. \qquad (3.2.8)$$

It follows therefore from (3.2.7), (3.2.8), and the homotopy invariance property that there exist $u_r \in P \cap S_r$ and $0 < \mu_r < 1$ such that $\mu_r \eta_0^{-1} A u_r = u_r$. Hence (3.2.2) holds for $\lambda_r = \eta_0 \mu_r^{-1} > \eta_0$.

If $-\infty < f'_+(0) < 0$, we can apply the result obtained above to the function $[-f(u)]$ and get the following conclusion: there exist $u_r \in P \cap S_r$ and $\lambda_r^* \geq \eta_0$ such that $\lambda_r^* u_r = -Au_r$ and hence (3.2.2) holds for $\lambda_r = -\lambda_r^*$.

When (H_4) is satisfied, the proof is similar. In this case we only need to change the definition of function ψ as follows:

$$\psi(x) = \begin{cases} 1, & \text{for } x \in D_2 \cup D_2^*; \\ 0, & \text{for } x \in G \quad (D_1 \cup D_1^*); \\ \text{number between } 0 \text{ and } 1 \text{ such that } \psi \text{ is continuous} \\ \qquad \text{on } G, \text{ for } x \in (D_1 \setminus D_2) \cup (D_1^* \setminus D_2^*), \end{cases}$$

where $D_1 = \{x \in R^n \mid |x - x_0| \leq \delta_1\} \subset G$ and $D_1^* = \{x \in R^n \mid |x - y_0| \leq \delta_1\} \subset G$ such that $k(x,y) \geq \tau$ $(\tau > 0)$ for $(x,y) \in D_1 \times D_1^*$ and $D_2 = \{x \in R^n \mid |x - x_0| \leq \delta_1/2\}$, $D_2^* = \{x \in R^n \mid |x - y_0| \leq \delta_1/2\}$.

Finally, in the case that (H_5) is satisfied, the proof is also similar. Although $k(x,y)$ may be discontinuous at points (x,y) with $x = y$, condition (H_5) will guarantee that the linear integral operator

$$ku(x) = \int_G k(x,y)u(y)dy$$

Is completely continuous from $C(G)$ into $C(G)$ (see, for example, Guo [4]), and therefore A is completely continuous from $P \cap \overline{B}_{\varepsilon_0}$ into P. The construction of the function ψ is same as in the case that (H_4). □

Notice that condition (ii) of Theorem 3.2.1 is easy to verify. For example, the polynomial

$$f(u) = \sum_{i=1}^{m} a_i u^i, \quad a_1 \neq 0 \tag{3.2.9}$$

satisfies (ii), since $f'_+(0) = a_1 \neq 0$.

Theorem 3.2.2. Suppose that (i) the kernel $k(x,y)$ satisfies (H_2) or (H_4) or (H_5) and (ii') the function $f(u)$ is continuous on $0 \leq u \leq \delta$ $(\delta > 0)$, $f(0) = 0$ and $f'_+(0) = +\infty$ or $-\infty$. Then, for any $N > 0$ there exists $\varepsilon_N > 0$ such that equation (3.2.1) has a nontrivial and nonnegative continuous solution $u_r(x)$ with $\|u_r\| = r$ and $|\lambda_r| \geq N$ for any $0 < r < \varepsilon_N$, i.e.,

$$\lambda_r u_r(x) = Au_r(x),\ u_r(x) \in C(G),\ u_r(x) \geq 0,\ \|u_r\| = r,\ |\lambda_r| \geq N. \tag{3.2.10}$$

Proof. As in the proof of Theorem 3.2.1, we need only prove Theorem 3.2.2 in the case that (H_2) is satisfied. We first assume that $f'_+(0) = +\infty$. For any given $N > 0$, there exists $0 < \varepsilon_N \leq \delta$ such that

$$f(u) \geq \sigma^* u,\ \forall\ 0 \leq u \leq \varepsilon_N, \tag{3.2.11}$$

where $\sigma^* = N(\tau \text{ mes } D_2)^{-1}$, τ and D_2 are defined by (3.2.4). For any $0 < r < \varepsilon_N$, we proceed in the same way as in the proof of Theorem 3.2.1: replacing (3.2.5) we may assume that

$$u - N^{-1}Au = \theta,\ \forall\ u \in P \cap S_r \tag{3.2.12}$$

and replacing (3.2.6) we can prove

$$u - N^{-1}Au = t\psi,\quad \forall\, u \in P \cap S_r,\quad t \geq 0. \tag{3.2.13}$$

Hence $i(N^{-1}A, P \cap B_r, P) = 0$. Observing $i(\theta, P \cap B_r, P) = 1$, we can show easily that there exist $u_r \in P \cap S_r$ and $0 < \mu_r < 1$ such that $\mu_r N^{-1}Au_r = u_r$ and hence (3.2.10) holds with $\lambda_r = N\mu_r^{-1} > N$.

In the case that $f'_+(0) = -$, we can reduce it to the case discussed above by considering $[-f(u)]$ instead of $f(u)$.

Notice, if

$$\lim_{u\to+0} f'(u) = +\infty \text{ or } -\infty, \tag{3.2.14}$$

then, by the mean value theorem, it is easy to show that $f'_+(0) = +\infty$ or $-\infty$. We can point out easily some elementary functions, which satisfy condition (3.2.14); for example,

$$f(u) = \sqrt[3]{u} - 2u^2 , \tag{3.2.15}$$

$$f(u) = \ln(1 - \sqrt{u}). \tag{3.2.16}$$

Theorem 3.2.3. Suppose that (i) the kernel $k(x,y)$ satisfies (H_2) or (H_4) or (H_5) and (ii*) the function $f(u)$ is nonnegative and continuous for $0 \leq u \leq +\infty$, $f(0) = 0$, $f'_+(0) = 0$ and

$$\lim_{u\to+\infty} \frac{f(u)}{u} = 0, \tag{3.2.17}$$

$$0 < \varliminf_{u\to+\infty} f(u) < +\infty . \tag{3.2.18}$$

Then, for arbitrary given $N > 0$, there exists $\delta_N > 0$ such that equation (3.2.1) has at least two nontrivial nonnegative continuous solutions $u_\lambda^{(1)}(x)$ and $u_\lambda^{(2)}(x)$ for any $0 < \lambda < \delta_N$; moreover, $\max_{x\, G} u_\lambda^{(1)}(x) > N$.

Proof. As in the proof of Theorem 3.2.1, we need only to prove this theorem in the case that (H_2) is satisfied. For given $N > 0$, by virtue of (3.2.18) there exists $\eta > 0$ and $a > N$ such that

$$f(u) \geq \eta, \quad u \geq a.$$

Letting $\delta_N = \tau\eta a^{-1}$ mes D_1, where τ and D_1 are defined by (3.2.4), we prove that this δ_N is required. Let $0 < \lambda < \delta_N$. Obviously, operator $1/\lambda\, A$ is completely continuous from P into P, where

$P = \{u \in C(G) \mid u(x) \geq 0\}$ is the cone as in the proof of Theorem 3.2.1. On account of (3.2.17) there exists $b > 0$ such that

$$f(u) \leq \frac{\lambda}{2\|K\|} u, \quad u \geq b,$$

where $\|K\|$ denotes the norm of the linear integral operator

$$Ku(x) = \int_G k(x,y)u(y)dy,$$

which maps $C(G)$ into $C(G)$. Hence, letting $\beta = \max_{0 \leq u \leq b} f(u)$, we find

$$0 \leq f(u) \leq \frac{\lambda u}{2\|K\|} + \beta, \quad u \geq 0. \tag{3.2.20}$$

We choose

$$R > \max\left\{a, \frac{2\beta\|K\|}{\lambda}\right\}. \tag{3.2.21}$$

Because of $f(0) = 0$ and $f'_+(0)$ there exists $0 < r < a$ such that

$$0 \leq f(u) \leq \frac{\lambda u}{2\|K\|}, \quad 0 \leq u \leq r. \tag{3.2.22}$$

Now, we consider the following three bounded open convex sets of P: $U_1 = \{u \in P \mid \|u\| < r\}$, $U_2 = \{u \in P \mid \|u\| < R$ and $\min_{x \in D_1} u(x) > a\}$, and $U_3 = \{u \in P \mid \|u\| < R\}$. Obviously,

$$U_1 \subset U_3, \quad U_2 \subset U_3, \quad U_1 \cap U_2 = \phi \tag{3.2.23}$$

and $\overline{U}_1 = \{u \in P \mid \|u\| \leq r\}$, $\overline{U}_2 = \{u \in P \mid \|u\| \leq R$ and $\min_{x \in D_1} u(x) \geq a\}$, $\overline{U}_3 = \{u \in P \mid \|u\| \leq R\}$. For $u \in \overline{U}_1$, from (3.2.22) we have

$$\left\|\frac{1}{\lambda}Au\right\| \leq \frac{1}{\lambda}\|K\| \cdot \frac{\lambda}{2\|K\|} \cdot \|u\| = \frac{1}{2}\|u\| \leq \frac{r}{2} < r,$$

which implies

$$\frac{1}{\lambda}A(\overline{U}_1) \subset U_1 . \tag{3.2.24}$$

For $u \in \overline{U}_3$, by (3.2.20) and (3.2.21) we find

$$\|\frac{1}{\lambda}Au\| \leq \frac{1}{\lambda}\|K\| \cdot (\frac{\lambda}{2\|K\|} \cdot \|u\| + \beta) \leq \frac{R}{2} + \frac{\beta}{\lambda}\|K\| < R ,$$

and hence

$$\frac{1}{\lambda}A(\overline{U}_3) \subset U_3 . \tag{3.2.25}$$

When $u \in \overline{U}_2$, we have $\|u\| \leq R$ and $\min_{x \in D_1} u(x) \geq a$ and, by (3.2.25), $\|1/\lambda\ Au\| < R$. Moreover, observing (3.2.19) and (3.2.4), we get

$$\min_{x \in D_1} \frac{1}{\lambda}Au(x) \geq \frac{1}{\lambda}\int_{D_1} k(x,y)f[u(y)]dy \geq \frac{\tau\eta}{\lambda} \text{ mes } D_1 = \frac{a\delta_N}{\lambda} > a.$$

Consequently

$$\frac{1}{\lambda}A(\overline{U}_2) \subset U_2 . \tag{3.2.26}$$

It follows from (3.2.24), (3.2.25), (3.2.26), and Lemma 2.4.1 that

$$i(\frac{1}{\lambda} A, U_k, P) = 1, \quad k = 1,2,3 . \tag{3.2.27}$$

Thus, $1/\lambda\ A$ has a fixed point $u_\lambda^{(1)}$ in U_2. It is clear that

$$\max_{x \in G} u_\lambda^{(1)}(x) \geq \min_{x \in D_1} u_\lambda^{(1)}(x) > a > N.$$

On the other hand, from (3.2.27) we see

$$i(\frac{1}{\lambda}A, U_3 \setminus (\overline{U}_1 \cup \overline{U}_2), P) = i(\frac{1}{\lambda}A, U_3, P) - i(\frac{1}{\lambda}A, U_1, P) - i(\frac{1}{\lambda}A, U_2, P) = -1 \neq 0,$$

and hence $1/\lambda\ A$ has a fixed point $u_\lambda^{(2)}$ in $U_3 \setminus (\overline{U}_1 \cup \overline{U}_2)$, and our theorem is proved. □

It is easy to give some elementary functions which satisfy condition (ii*) of Theorem 3.2.3. For example,

$$f(u) = \sqrt{u} \ln(1 + u), \quad \text{and} \tag{3.2.28}$$

$$f(u) = \frac{u^2}{1 + u^2} + 2 \sin^2 u . \tag{3.2.29}$$

Also, from the proof of Theorem 3.2.3 it is clear that (3.2.26) is still true when condition $f'_+(0) = 0$ is realized, and therefore we get the following.

Theorem 3.2.4. Suppose that (i) the kernel $k(x,y)$ satisfies (H_2) or (H_4) or (H_5) and (ii_*) the function $f(u)$ is nonnegative and continuous for $0 \leq u < +\infty$, $f(0) = 0$ and (3.2.17), (3.2.18) hold. Then, for arbitrary given $N > 0$, there exists $\delta_N > 0$ such that equation (3.2.1) has a nontrivial and nonnegative continuous solution $u_\lambda(x)$ with $\max_{x \in G} u_\lambda(x) > N$ for any $0 < \lambda < \delta_N$.

Now, we apply Theorems 3.2.1-3.2.4 to the following two two-point boundary value problems:

$$\begin{cases} -\lambda x'' = f(x), & 0 \leq t \leq 1; \\ x(0) = x(1) = 0 \end{cases} \tag{3.2.30}$$

and

$$\begin{cases} -\lambda x'' = f(x), & 0 \leq t \leq 1; \\ x(0) = x'(1) = 0 \end{cases} \tag{3.2.31}$$

and the following Dirichlet problem:

$$\begin{cases} \lambda Lu = f(u), & x \in \Omega; \\ u\big|_{\partial\Omega} = 0, \end{cases} \tag{3.2.32}$$

where L is a uniformly elliptic differential operator as described in Example 2.1.2. Since the Green functions corresponding to problems (3.2.30) and (3.2.31) satisfy the condition (H_2), and the Green function of L with Dirichlet boundary condition satisfies condition (H_5) (see (2.1.37)), Theorems 3.2.1-3.2.4 imply Theorems 3.2.5 and 3.2.6:

Theorem 3.2.5. Let $f(x)$ be continuous on $0 \le x \le \delta$ for some $\delta > 0$ and $f(0) = 0$. Then the following conclusions hold:

(a) if $f'_+(0) \neq 0$, then there exist $\varepsilon_0 > 0$ and $\eta_0 > 0$ such that for any $0 < r < \varepsilon_0$ problems (3.2.30) and (3.2.31) have nontrivial solutions $\lambda_r^{(1)}$, $x_r^{(1)}(t)$ and $\lambda_r^{(2)}$, $x_r^{(2)}(t)$, respectively, which satisfy $x_r^{(i)}(t) \in C^2[0,1]$, $\|x_r^{(i)}\|_c = r$, $|\lambda_r^{(i)}| \ge \eta_0$ $(i = 1,2)$ and $x_r^{(1)}(t) > 0$ for $0 < t < 1$, $x_r^{(2)}(t) > 0$ for $0 < t \le 1$;

(b) if $f'_+(0) = \pm\infty$, then for any given $N > 0$ there exists $\varepsilon_N > 0$ such that for any $0 < r < \varepsilon_N$ problems (3.2.30) and (3.2.31) have nontrivial solutions $\lambda_r^{(1)}$, $x_r^{(1)}(t)$ and $\lambda_r^{(2)}$, $x_r^{(2)}(t)$, respectively, which satisfy $x_r^{(i)}(t) \in C^2[0,1]$, $\|x_r^{(i)}\| = r$, $|\lambda_r^{(i)}| \ge N$ $(i = 1,2)$ and $x_r^{(1)}(t) > 0$ for $0 < t < 1$, $x_r^{(2)}(t) > 0$ for $0 < t \le 1$;

(c) if $f'_+(0) = 0$ and $f(x)$ is nonnegative and continuous for $0 \le x < +\infty$ such that (3.2.17) and (3.2.18) hold, then for arbitrary given $N > 0$, there exists $\delta_N > 0$ such that for any $0 < \lambda < \delta_N$ problems (3.2.30) and (3.2.31) have at least two nontrivial solutions $x_\lambda^{(1)}(t)$, $x_\lambda^{(2)}(t)$ and $y_\lambda^{(1)}(t)$, $y_\lambda^{(2)}(t)$, respectively, which satisfy $x_\lambda^{(i)}(t) \in C^2[0,1]$, $y_\lambda^{(i)}(t) \in C^2[0,1]$, $x_\lambda^{(i)}(t) > 0$ for $0 < t < 1$, $y_\lambda^{(i)}(t) > 0$ for $0 < t \le 1$ $(i = 1,2)$ and $\max_{0\le t\le 1} x_\lambda^{(1)}(t) > N$, $\max_{0\le t\le 1} y_\lambda^{(1)}(t) > N$.

Theorem 3.2.6. Let $f(u) \in \overline{C^\alpha}([0,\delta])$ for some $0 < \alpha < 1$ and $\delta > 0$ and $f(0) = 0$. Then the following conclusions hold:

(a) if $f'_+(0) \neq 0$, then there exist $\varepsilon_0 > 0$ and $\eta_0 > 0$ such that for any $0 < r < \varepsilon_0$ problem (3.2.32) has nontrivial solution λ_r, $u_r(x)$ with $u_r(x) \in C^{2+\gamma}(\overline{\Omega})$, where $\gamma = \min\{\mu,\alpha\}$, $\|u_r\|_C = r$, $u_r(x) > 0$ for $x \in \Omega$ and $|\lambda_r| \ge \eta_0$;

(b) if $f'_+(0) = \pm\infty$, then for any given $N > 0$ there exists $\varepsilon_N > 0$ such that for any $0 < r < \varepsilon_N$ problem (3.2.32) has nontrivial solution λ_r,

$u_r(x)$ with $u_r(x) \in C^{2+\gamma}(\bar{\Omega})$, where $\gamma = \min\{\mu,\alpha\}$, $\|u_r\|_C = r$, $u_r(x) > 0$ for $x \in \Omega$ and $|\lambda_r| \geq N$;

(c) if $f'_+(0) = 0$ and $f(u)$ is nonnegative for $0 \leq u < +\infty$ such that $f(u) \in C^{\alpha}([0,b])$ for all $b > 0$ and (3.2.17) and (3.2.18) hold, then for arbitrary given $N > 0$, there exists $\delta_N > 0$ such that problem (3.2.32) has at least two nontrivial solutions $u_\lambda^{(i)}(x) \in C^{2+\gamma}(\bar{\Omega})$ for any $0 < \lambda < \delta_N$, where $\gamma = \min\{\mu,\alpha\}$ and $u_\lambda^{(i)}(x) > 0$ for $x \in \Omega$ $(i = 1,2)$ and $\max_{x\in\bar{\Omega}} u_\lambda^{(1)}(x) > N$.

Notice that for function (3.2.9) we have $f'_+(0) = 0$ and $f(u) \in C^{\alpha}([0,\delta])$ for any $0 < \alpha < 1$, and hence, by Theorem 3.2.6(a), $u_r(x) \in C^{2+\mu}(\bar{\Omega})$. For function (3.2.15) we have $f'_+(0) = +\infty$ and $f(u) \in C^{1/3}([0,\delta])$ and for function (3.2.16) we find $f'_+(0) = -\infty$ and $f(u) \in C^{1/2}([0,\delta])$ for $0 < \delta < 1$ and therefore by virtue of Theorem 3.2.6(b), $u_r(x) \in C^{2+\gamma}(\bar{\Omega})$ with $\xi = \min\{\mu,1/3\}$ and $u_r(x) \in C^{2+\xi}(\bar{\Omega})$ with $\xi = \min\{\mu,1/2\}$, respectively. Finally, function (3.2.28) belongs to $C^{1/2}([0,b])$ for any $b > 0$ and function (3.2.29) belongs to $C^{\alpha}([0,b])$ for any $0 < \alpha < 1$ and any $b > 0$. They also satisfy $f'_+(0) = 0$ and hence, by Theorem 3.2.6(c), the two positive solutions of problem (3.2.32) corresponding to functions (3.2.28) and (3.2.29) belong to $C^{2+\gamma}(\bar{\Omega})$ $(\gamma = \min\{\mu,1/2\})$ and $C^{2+\mu}(\bar{\Omega})$, respectively.

Theorem 3.2.7. Suppose that (a) the kernel $k(x,y)$ satisfies (H_1) and (b) the function $f(u)$ is continuous on $0 \leq u < \delta$ $(\delta > 0)$, $f(0) = 0$ and $f(u) \neq 0$ for $0 < u < \delta$. Then for any $0 < r < \delta$ equation (3.2.1) has a nontrivial and nonnegative continuous solution $u_r(x)$ with $\|u_r\| = r$ and $\lambda_r \neq 0$, i.e.,

$$\lambda_r u_r(x) = Au_r(x),\ u_r(x) \in C(G),\ u_r(x) \geq 0,\ \|u_r\| = r,\ \lambda_r \neq 0. \quad (3.2.33)$$

<u>Proof</u>. We first assume that $f(u) > 0$ for $0 < u < \delta$. Let $0 < r < \delta$ be given. Denoting $B_r = \{u \in C(G) \mid \|u\| < r\}$, $S_r = \{u \in C(G) \mid \|u\| = r\}$ and

$$P = \{u \in C(G) \mid u(x) \geq 0 \text{ and } \min_{x \in G_0} u(x) \geq \varepsilon_0 \|u\|\},$$

we see that P is a cone of $E = C(G)$ and $A : P \cap \overline{B}_r \to P$ is completely continuous. As in the proof of Theorem 3.1.1, we can get

$$A(P \cap \overline{B}_r) \subset P. \tag{3.2.34}$$

Choose ε_1 sufficiently small such that $0 < \varepsilon_1 < r\varepsilon_0$ and, for any $u \in P \cap S_r$, let $G_1 = \{x \in G_0 \mid u(x) \geq \varepsilon_1\}$. We have

$$\int_{G_0} u(x)dx = \int_{G_1} u(x)dx + \int_{G_0 \setminus G_1} u(x)dx \leq r \text{ mes } G_1$$
$$+ \varepsilon_1(\text{mes } G_0 - \text{mes } G_1).$$

On the other hand,

$$\int_{G_0} u(x)dx \geq \varepsilon_0 \|u\| \text{mes } G_0 = \varepsilon_0 r \text{ mes } G_0,$$

and hence

$$\text{mes } G_1 \geq \frac{(\varepsilon_0 r - \varepsilon_1)\text{mes } G_0}{r - \varepsilon_1} = \varepsilon_2,$$

where $\varepsilon_2 > 0$ is obviously independent of u. Now, let

$$\tau = \min_{(x,y) \in G_0 \times G_0} k(x,y) > 0 \quad \text{and} \quad = \min_{\varepsilon_1 \leq u \leq r} f(u) > 0.$$

Then, for $x \in G_0$, we have

$$Au(x) \geq \int_{G_1} k(x,y) f[u(y)]dy \geq \tau\eta \text{ mes } G_1 \geq \tau\eta\varepsilon_2 ,$$

and so $\|Au\| \geq \tau\eta\varepsilon_2$, which implies

$$\inf_{u \in P \cap S_r} \|Au\| \geq \tau\eta\varepsilon_2 > 0. \tag{3.2.35}$$

It follows from (3.2.35) and Theorem 2.3.6 that there exist $u_r \in P \cap S_r$ and $\lambda_r > 0$ such that $Au_r = \lambda_r u_r$, i.e., (3.2.33) holds.

In case that $f(u) < 0$ for $0 < u < \delta$, we can apply the conclusion obtained above to the function $[-f(u)]$ and assert that there exist $u_r \in P \cap S_r$ and $\lambda_r' > 0$ such that $-Au_r = \lambda_r' u_r$; hence (3.2.33) also holds for $\lambda_r = -\lambda_r' < 0$.

Theorem 3.2.8. Suppose that (a) the kernel $k(x,y)$ satisfies (H_1) and (b') the function $f(u)$ is continuous and nonnegative or nonpositive on $0 \leq u < +\infty$, $f(0) = 0$ and $f(u) \neq 0$ for $u > R$ $(R > 0)$. Then for any $r > R\varepsilon_0^{-1}$ equation (3.2.1) has a nontrivial and nonnegative continuous solution $u_r(x)$ with $\|u_r\| = r$ and $\lambda_r \neq 0$, i.e., (3.2.33) holds.

Proof. As in the proof of Theorem 3.2.7, we may assume that $f(u) \geq 0$ on $u \geq 0$ and $f(u) > 0$ for $u > R$. Let $r > R\varepsilon_0^{-1}$ be given. Using the same method we can get (3.2.34). Since $r > R\varepsilon_0^{-1}$, we can choose ε_1 such that $R < \varepsilon_1 < r\varepsilon_0$. As in the proof of Theorem 3.2.7, we obtain (3.2.35), and so there exists $u_r \in P \cap S_r$ such that (3.2.33) holds. □

By the discussion before Theorem 3.1.4, we see that the Green functions of (3.1.30) and (3.1.31) satisfy condition (H_1) with $\varepsilon_0 = \alpha(1 - \beta)$ and $\varepsilon_0 = \alpha$, respectively, where α and β may be any numbers satisfying $0 < \alpha < \beta < 1$. Observing

$$\sup_{0<\alpha<\beta<1} \alpha(1 - \beta) = \frac{1}{4} \quad \text{and} \quad \sup_{0<\alpha<\beta<1} \alpha = 1,$$

we obtain the following two theorems by virtue of Theorems 3.2.7 and 3.2.8.

Theorem 3.2.9. (i) if $f(x)$ is continuous on $0 \leq x < \delta$ for some $\delta > 0$, $f(0) = 0$ and $f(x) \neq 0$ for $0 < x < \delta$, then for any $0 < r < \delta$ the problem (3.2.30) has a nontrivial solution λ_r, $x_r(t)$ such that

$$x_r(t) \in C^2[0,1],\ x_r(t) > 0 \quad (0 < t < 1), \quad \max_{0 \le t \le 1} x_r(t) = r, \quad \lambda_r \ne 0. \tag{3.2.36}$$

(ii) If $f(x)$ is continuous and nonnegative or nonpositive on $0 \le x < +\infty$, $f(0) = 0$ and $f(x) \ne 0$ for $x > R$ $(R > 0)$, then for any $r > 4R$ the problem (3.2.30) has a nontrivial solution λ_r, $x_r(t)$ such that (3.2.36) holds.

Theorem 3.2.10. (i') If $f(x)$ is continuous on $0 \le x < \delta$ for some $\delta > 0$, $f(0) = 0$ and $f(x) \ne 0$ for $0 < x < \delta$, then for any $0 < r < \delta$ the problem (3.2.31) has a nontrivial solution λ_r, $x_r(t)$ such that

$$x_r(t) \in C^2[0,1],\ x_r(t) > 0 \quad (0 < t \le 1), \quad \max_{0 \le t \le 1} x_r(t) = r,\ \lambda_r \ne 0. \tag{3.2.37}$$

(ii') If $f(x)$ is continuous and nonnegative or nonpositive on $0 \le x < +\infty$, $f(0) = 0$ and $f(x) \ne 0$ for $x > R$ $(R > 0)$, then for any $r > R$ the problem (3.2.31) has a nontrivial solution λ_r, $x_r(t)$ such that (3.2.37) holds.

Notice that the conditions on $f(u)$ in Theorems 3.2.7-3.2.10 are very weak and easy to verify.

3.3 *Some Nonlinear Integral Equations Arising in Science.*

(a) A Nonlinear Integral Equation Arising in the Theory of Neutron Transport.

Consider an unbounded slab nuclear reactor between two planes $x = -a$ and $x = a$. The following nonlinear integro-differential equation occurs in the theory of neutron transport (see Pazy and Rabinowitz [1]):

$$\mu \frac{\partial u}{\partial x} - \sigma u + \frac{c\sigma}{2} \int_{-1}^{1} u(x,\mu')d\mu' = \sigma F(\frac{1}{2} \int_{-1}^{1} u(x,\mu')d\mu'), \tag{3.3.1}$$

where

$$F(z) = \sum_{k=2}^{N} c_k[(1 - z)^k - 1 + kz], \quad c_k \geq 0, \quad \sum_{k=0}^{N} c_k = 1, \quad \sum_{k=1}^{N} kc_k = c,$$

and $c_2, c_3, \ldots, c_N$ are not all equal to zero. The boundary condition is

$$u(a,\mu) = 0 \text{ for } \mu > 0, \quad u(-a,\mu) = 0 \text{ for } \mu < 0. \tag{3.3.2}$$

The solution $u(x,\mu)$ of equation (3.3.1) represents the probability that a neutron injected into the reactor at a point x with direction ($\mu = \cos(\ell, x)$) will produce a persisting chain reaction, and hence $0 \leq u(x,\mu) \leq 1$. In a collision of a neutron with an atom of the reactor, the neutron has a probability c_0 of being absorbed and a probability c_k $(1 \leq k \leq N)$ of being replaced instantaneously by k new neutrons starting from the point of the collision and having random directions. Thus, the parameter c is the average number of neutrons emerging from a collision. In addition, σ^{-1} is the mean free path of a neutron between two collisions.

Writing (3.3.1) in the following form

$$\begin{cases} -\mu e^{\sigma x/\mu}\left(\dfrac{\partial}{\partial x} e^{-\sigma x/\mu} u\right) = \sigma G(\phi(x)), & \mu \neq 0; \\ u(x,0) = G(\phi(x)), & \mu = 0, \end{cases} \tag{3.3.3}$$

where

$$\phi(x) = \frac{1}{2}\int_{-1}^{1} u(x,\mu')d\mu', \tag{3.3.4}$$

$$G(z) = cz - F(z) = 1 - \sum_{k=0}^{N} c_k(1 - z)^k, \tag{3.3.5}$$

and integrating and using the boundary condition (3.3.2), we get

$$
u(x,\mu) = \begin{cases} \int_x^a e^{\sigma(x-y)/\mu} \cdot \frac{\sigma}{\mu} G(\phi(y))dy, & \mu > 0; \\ G(\phi(x)), & \mu = 0; \\ -\int_{-a}^x e^{\sigma(x-y)/\mu} \cdot \frac{\sigma}{\mu} G(\phi(y))dy, & \mu < 0. \end{cases} \tag{3.3.6}
$$

A further integration of (3.3.6) with respect to μ gives, after interchanging the order of integration and making the change of variable $t = \sigma/\mu$,

$$
\phi(x) = \int_{-a}^{a} E(|x - y|)G(\phi(y))dy, \tag{3.3.7}
$$

where

$$
E(z) = \frac{\sigma}{2} \int_{\sigma}^{+\infty} \frac{e^{-tz}}{t} dt. \tag{3.3.8}
$$

Conversely, if $\phi(x)$ is a continuous solution of (3.3.7), then use of (3.3.6) leads to a C^1 solution of (3.3.1)-(3.3.2). Moreover, $0 \leq u(x,\mu) \leq 1$ if and only if $0 \leq \phi(x) \leq 1$. Thus, we have reduced the problem (3.3.1)-)3.3.2) to an equivalent Hammerstein nonlinear integral equation (3.3.7).

Now, consider the Hammerstein nonlinear integral operator

$$
A\phi = \int_{-a}^{a} E(|x - y|)G(\phi(y))dy \tag{3.3.9}
$$

and the corresponding linear integral operator

$$
B\phi = \int_{-a}^{a} E(|x - y|)\phi(y)dy . \tag{3.3.10}
$$

Let $D = \{\phi \in C[-a,a] \mid 0 \leq \phi(x) \leq 1\}$. It is easy to get the following Lemma (see Pazy and Rabinowitz [1] and Guo [1], [13]).

<u>Lemma 3.3.1</u>. (i) B is completely continuous from $C[-a,a]$ into $C[-a,a]$ and the norm

$$\|B\| = 1 - \sigma \int_a^{+\infty} \frac{e^{-ta}}{t^2} dt = 1 - e^{-\sigma a} + 2aE(a) ;$$

(ii) A is completely continuous from D into D;

(iii) B has a unique nonnegative continuous normalized characteristic function $\psi^*(x)$. Moreover, $\psi^* = c^*B\psi^*$, $c^* > 1$, $\psi^*(x) > 0$ for $x \in [-a,a]$, $\|\psi^*\| = 1$ and c^* is a simply characteristic value of B in such a way that $|\bar{c}| > c^*$ for any other characteristic value $\bar{c}$ of B. In addition, $\psi^*(x)$ is an even function, i.e.,

$$\psi^*(-x) = \psi^*(x), \quad -a \leq x \leq a.$$

Theorem 3.3.1. If $0 < c \leq c^*$, then equation (3.3.7) has only the trivial solution $\phi(x) \equiv 0$ in D.

Proof. Suppose that equation (3.3.7) has a nontrivial solution $\phi^*(x) \not\equiv 0$ in D. Since $G_z'(z) = \sum_{k=1}^{N} kc_k(1 - z)^{k-1}$ is strictly decreasing in $0 \leq z \leq 1$, we have

$$\phi^*(x) = \int_{-a}^{a} E(|x - y|)G(\phi^*(y))dy = \int_{-a}^{a} E(|x - y|)G_z'(\theta(y)\phi^*(y))\phi^*(y)dy$$

$$< \int_{-a}^{a} E(|x - y|)G_z'(0)\phi^*(y)dy = c \int_{-a}^{a} E(|x - y|)\phi^*(y)dy = cB\phi^*(x),$$

where $0 < \theta(y) < 1$. Hence

$$\max_{0 \leq x \leq 1} \frac{\phi^*(x)}{B\phi^*(x)} = c_1 < c \leq c^* ,$$

and so

$$\phi^* \leq c_1 B\phi^*, \quad 0 < c_1 < c^*. \tag{3.3.11}$$

Now, let

$$B_n\phi = B\phi + \frac{\phi^*}{\eta}, \quad \bar{B}_n\phi = \frac{B_n\phi}{\|B_n\phi\|} \quad (n = 1,2,3,\ldots) . \tag{3.3.12}$$

Since $\phi \in D$ implies $\|B_n\phi\| \geq \|\phi^*\|/n > 0$, we see that $\overline{B}_n$ is a completely continuous operator from D into D. Observing that D is a bounded closed convex set in $C[-a,a]$, we know by Schauder fixed point theorem that there exist $\phi_n \in D$ such that $\overline{B}_n\phi_n = \phi_n$ $(n = 1,2,3,\ldots)$. Obviously,

$$\|\phi_n\| = 1, \quad B\phi_n + \frac{\phi^*}{n} = \lambda_n\phi_n \quad (n = 1,2,3,\ldots), \tag{3.3.13}$$

where

$$\frac{\|\phi^*\|}{n} \leq \lambda_n = \|B_n\phi_n\| \leq \|B\| + \frac{\|\phi^*\|}{n} \quad (n = 1,2,3,\ldots). \tag{3.3.14}$$

Let $t^* = \sup\{t > 0 \mid \phi_n \geq t\phi^*\}$. It follows from (3.3.13) that $0 < 1/n\lambda_n \leq t^* < +\infty$ and $\phi_n \geq t^*\phi^*$. Hence, by virtue of (3.3.13) and (3.3.11) we find

$$\phi_n \geq \frac{1}{\lambda_n} B\phi_n \geq \frac{1}{\lambda_n} B(t^*\phi^*) = \frac{t^*}{\lambda_n} B\phi^* \geq \frac{t^*}{c_1\lambda_n}\phi^*,$$

and therefore $t^*(c_1\lambda_n)^{-1} \leq t^*$, i.e.,

$$\lambda_n \geq \frac{1}{c_1} \quad (n = 1,2,3,\ldots). \tag{3.3.15}$$

By the complete continuity of B and (3.3.14) and (3.3.15) we see that there exists a subsequence $\{n_k\}$ of $\{n\}$ such that $B\phi_{n_k} \to \psi_0$ and $\lambda_{n_k} \to \lambda_0$, where $c_1^{-1} \leq \lambda_0 \leq \|B\|$. Thus, by (3.3.13) we get $\phi_{n_k} \to \lambda_0^{-1}\psi_0 = \phi_0$ with $\|\phi_0\| = 1$, and so $B\phi_0 = \lambda_0\phi_0$, i.e., $\phi_0 = \mu_0 B\phi_0$, $\mu_0 = \lambda_0^{-1}$, $0 < \mu_0 \leq c_1$. Hence, μ_0 is a characteristic value of B. But by (3.3.11) we know $0 < \mu_0 < c^*$, which contradicts Lemma 3.3.1 (iii).□

Lemma 3.3.2. If $c > c^*$, then equation (3.3.7) has a minimal nontrivial solution $u(x)$ and a maximal nontrivial solution $v(x)$ in D; moreover,

$$0 < u(x) \leq v(x) < 1, \quad -a \leq x \leq a. \tag{3.3.16}$$

Proof. Since $G(z)$ is strictly increasing in $0 \leq z \leq 1$, operator A (see (3.3.9)) is increasing from D into D. Now, let

$$v_0(x) \equiv 1, \quad v_n(x) = Av_{n-1}(x) \quad (n = 1,2,3,\ldots). \tag{3.3.17}$$

Since

$$\begin{aligned} v_1(x) &= \int_{-a}^{a} E(|x - y|)G(1)dy \\ &= \frac{1}{2}\sigma(1 - c_0)\int_{-a}^{a} dy \int_{\sigma}^{+\infty} \frac{e^{-t|x-y|}}{t} dt \\ &= \frac{1}{2}\sigma(1 - c_0)\int_{\sigma}^{+\infty} dt \int_{-a}^{a} \frac{e^{-t|x-y|}}{t} dy \\ &= (1 - c_0)\cdot\left\{1 - \sigma\int_{\sigma}^{+\infty} \frac{e^{-ta}}{t^2}\cosh(tx)dt\right\} < 1 = v_0(x), \end{aligned}$$

we have $v_0 \geq v_1 \geq v_2 \geq \ldots \geq \theta$, and therefore, by Theorem 2.1.1, there exists $v \in D$ such that $\|v_n - v\| \to 0$ as $n \to \infty$ and v is the maximal fixed point of A in D; moreover,

$$0 \leq v(x) \leq v_1(x) < 1, \ \forall \ -a \leq x \leq a. \tag{3.3.18}$$

On the other hand, choosing $\delta > 0$ sufficiently small such that

$$(1 - \delta)c > c^* \tag{3.3.19}$$

and observing the continuity of $G_z'(z)$ and $G_z'(0) = \sum_{k=1}^{N} kc_k = c$, we can select $\varepsilon_0 > 0$ such that

$$G_z'(z) \geq (1 - \delta)c, \ \forall \ 0 \leq z \leq \varepsilon_0. \tag{3.3.20}$$

Hence, for $0 < \varepsilon \leq \varepsilon_0$, we have

$$A(\varepsilon\psi^*(x)) = \int_{-a}^{a} E(|x - y|)G(\varepsilon\psi^*(y))dy$$

$$= \int_{-a}^{a} E(|x - y|)G_z'(\theta(y)\varepsilon\psi^*(y))\varepsilon\psi^*(y)dy$$

$$\geq \int_{-a}^{a} E(|x - y|)(1 - \delta)c\ \varepsilon\ \psi^*(y)dy$$

$$= \frac{(1 - \delta)c}{c^*}\ \psi^*(x) \geq \varepsilon\psi^*(x)\ . \qquad (3.3.21)$$

Now, let

$$u_{0,\varepsilon}(x) = \varepsilon\psi^*(x),\ u_{n,\varepsilon}(x) = Au_{n-1,\varepsilon}(x) \quad (n = 1,2,3,\ldots). \qquad (3.3.22)$$

By (3.3.21) we see

$$\varepsilon\psi^* = u_{0,\varepsilon} \leq u_{1,\varepsilon} \leq u_{2,\varepsilon} \leq \cdots \leq 1,$$

and hence, there exists $u_\varepsilon \in D$ such that $\|u_{n,\varepsilon} - u_\varepsilon\| \to 0$ as $n \to \infty$ and $Au_\varepsilon = u_\varepsilon$, $0 < u_\varepsilon(x) \leq v(x) < 1$, $\forall\ -a \leq x \leq a$.

In the following we prove that u_ε is independent of ε for $0 < \varepsilon \leq \varepsilon_0$. Indeed, assume $0 < \varepsilon_1 < \varepsilon_2 < \varepsilon_0$. Clearly $u_{\varepsilon_1} \leq u_{\varepsilon_2}$. Let $T = \{\varepsilon \mid \varepsilon_1 \leq \varepsilon \leq \varepsilon_2$ and $\varepsilon\psi^* \leq u_{\varepsilon_1}$. Since $\varepsilon_1 \in T$, T is not empty. Now, let $\varepsilon^* = \sup T$ and prove $\varepsilon^* = \varepsilon_2$. If $\varepsilon^* < \varepsilon_2$, then, letting $\varepsilon = \varepsilon^*$ in (3.3.21), we get

$$\varepsilon^*\psi^* \leq \frac{(1 - \delta)c\varepsilon^*}{c^*}\ \psi^* \leq A(\varepsilon^*\psi^*) \leq Au_{\varepsilon_1} = u_{\varepsilon_1}\ ,$$

which contradicts the definition of ε^* since $(1 - \delta)c\varepsilon^*/c^* > \varepsilon^*$. Hence $\varepsilon^* = \varepsilon_2$, $\varepsilon_2\psi^* \leq u_{\varepsilon_1}$, and so $u_{\varepsilon_2} \leq u_{\varepsilon_1}$. Thus, $u_{\varepsilon_1} = u_{\varepsilon_2}$ and u_ε is independent of ε for $0 < \varepsilon \leq \varepsilon_0$, which will be denoted by u.

Finally, we prove that u is the minimal nonzero fixed point of A in D. In fact, suppose $\phi^* \in D$ such that $\phi^* \neq \theta$ and $A\phi^* = \phi^*$. Then

$$\phi^*(x) = \int_{-a}^{a} E(|x-y|)G(\phi^*(y))dy \geq E(2a) \int_{-a}^{a} G(\phi^*(y))dy = \alpha > 0,$$

$$-a \leq x \leq a.$$

Choosing ε such that $0 < \varepsilon < \min\{\varepsilon_0, \alpha\}$, we see $\varepsilon\psi^* \leq \varepsilon < \alpha \leq \phi^*$, and so $u_{n,\varepsilon} \leq \phi^*$ for $n = 1,2,3,\ldots$. Taking limit we get $u = u_\varepsilon \leq \phi^*$, and therefore u is the minimal nonzero fixed point of A in D and

$$0 < u(x) \leq v(x) < 1, \ \forall \ -a \leq x \leq a. \ \square$$

<u>Theorem 3.3.2</u>. If $c \geq c^*$, then equation (3.3.7) has exactly one non-trivial solution $\phi^*(x)$ in D; moreover

$$0 < \phi^*(x) < 1, \ \forall \ -a \leq x \leq a \tag{3.3.23}$$

and, constructing successively the sequence

$$\phi_n(x) = \int_{-a}^{a} E(|x-y|)G(\phi_{n-1}(y))dy \quad (n = 1,2,3,\ldots) \tag{3.3.24}$$

for any initial not identically zero function $\phi_0(x) \in D$, we have

$$\|\phi_n - \phi^*\| = \max_{-a \leq x \leq a} |\phi_n(x) - \phi^*(x)| \to 0 \quad (n \to \infty). \tag{3.3.25}$$

<u>Proof</u>. By virtue of Lemma 3.3.2, in order to prove the uniqueness of non-trivial solutions of equation (3.3.7) in D it is sufficient to verify

$$u(x) \equiv v(x), \ \forall \ -a \leq x \leq a. \tag{3.3.26}$$

Now, we first prove

$$\frac{G(z_1)}{z_1} > \frac{G(z_2)}{z_2}, \ \forall \ 0 < z_1 < z_2 < 1. \tag{3.3.27}$$

In fact, $G'_z(z) = \sum_{k=1}^{N} kc_k(1-z)^{k-1}$ is strictly decreasing in $0 \leq z \leq 1$ and $G(0) = 0$. Consequently, for $0 < z < 1$, we have

$$\frac{d}{dz}\left(\frac{G(z)}{z}\right) = \frac{zG_z'(z) - G(z)}{z^2} = \frac{1}{z^2}\{zG_z'(z) - [G(z) - G(0)]\}$$

$$= \frac{1}{z^2}\{zG_z'(z) - G_z'(\theta z)z\} = \frac{1}{z}\{G_z'(z) - G_z'(\theta z)\} < 0 ,$$

where $0 < \theta < 1$. It follows therefore that $G(z)/z$ is a strictly decreasing function on $0 < z < 1$ and so (3.3.27) holds.

If $u(x)$ and $v(x)$ are not identical, then by (3.3.16) and (3.3.27) we have

$$\frac{G(u(x))}{u(x)} - \frac{G(v(x))}{v(x)} \geq 0, \quad \frac{G(u(x))}{u(x)} - \frac{G(v(x))}{v(x)} \not\equiv 0, \ \forall \ -a \leq x \leq a,$$

and so, by virtue of (3.3.16) and the continuity of functions $u(x)$, $v(x)$ and $G(z)$,

$$\int_{-a}^{a} v(x)u(x)\cdot\left[\frac{G(u(x))}{u(x)} - \frac{G(v(x))}{v(x)}\right]dx > 0.$$

But, on the other hand, we see that

$$\int_{-a}^{a} v(x)u(x)\cdot\left[\frac{G(u(x))}{u(x)} - \frac{G(v(x))}{v(x)}\right]dx$$

$$= \int_{-a}^{a} [v(x)G(u(x)) - u(x)G(v(x))]dx$$

$$= \int_{-a}^{a}\int_{-a}^{a} E(|x - y|)G(v(y))G(u(x))dxdy$$

$$- \int_{-a}^{a}\int_{-a}^{a} E(|x - y|)G(u(y))G(v(x))dxdy$$

$$= \int_{-a}^{a}\int_{-a}^{a} E(|x - y|)G(v(y))G(u(x))dxdy$$

$$- \int_{-a}^{a}\int_{-a}^{a} E(|y - x|)G(v(x))G(u(y))dxdy = 0,$$

which contradicts (3.3.28), and so (3.3.26) holds.

Finally, it remains to prove (3.3.25). From (3.3.24) we see

$$\phi_n = A\phi_{n-1} \quad (n = 1,2,3,\ldots), \quad \phi_0 \in D, \quad \phi_0 \neq \theta,$$

and so

$$1 \geq \phi_1(x) = \int_{-a}^{a} E(|x - y|)G(\phi_0(y))dy \geq E(2a) \int_{-a}^{a} G(\phi_0(y))dy = \beta > 0.$$

Now, choose ε sufficiently small such that $0 < \varepsilon < \min\{\varepsilon_0, \beta\}$, where ε_0 is defined by (3.3.20). Hence

$$\varepsilon\psi^* \leq \phi_1 \leq 1 = v_0.$$

Operating this inequality $n - 1$ times, we get

$$u_{n-1,\varepsilon} \leq \phi_n \leq v_{n-1} \quad (n = 1,2,3,\ldots), \tag{3.3.29}$$

where $\{v_n\}$ and $\{u_{n,\varepsilon}\}$ are the sequences (3.3.17) and (3.3.22), respectively. Since $u_{n,\varepsilon} \to u_\varepsilon = u$ and $v_n \to v$ as $n \to \infty$ and $u = v = \phi^*$, (3.3.25) follows by taking limit in (3.3.29).

Corollary 3.3.1. $\phi^*(x)$ is an even function, i.e.,

$$\phi^*(-x) = \phi^*(x), \quad \forall -a \leq x \leq a. \tag{3.3.30}$$

Proof. Letting $\psi(x) = \phi^*(-x)$, we see that

$$\psi(x) = \int_{-a}^{a} E(|-x - y|)G(\phi^*(y))dy$$

$$= \int_{-a}^{a} E(|-x + z|)G(\phi^*(-z))dz$$

$$= \int_{-a}^{a} E(|x - z|)G(\psi(z))dy \;,$$

and hence $\psi(x)$ is also a nontrivial solution of equation (3.3.7) in D. By uniqueness it must be $\psi(x) \equiv \phi^*(x)$, i.e., (3.3.30) holds.

Now, by the equivalence of problem (3.3.1)-(3.3.2) and equation (3.3.7) the following theorem follows immediately from Theorems 3.3.1 and 3.3.2.

Theorem 3.3.3. If $0 < c \leq c^*$, then problem (3.3.1)-(3.3.2) has only the trivial solution $u(x,\mu) \equiv 0$. If $c > c^*$, then it has exactly one nontrivial solution $u(x,\mu)$. Moreover,

$$0 < u(x,\mu) < 1, \ \forall \ -a < x < a, \quad -1 \leq \mu \leq 1,$$

$$u(x,\mu) = u(-x,-\mu), \ \forall \ -a \leq x \leq a, \quad -1 \leq \mu \leq 1,$$

and

$$\sup_{-a\leq x\leq a,-1\leq\mu\leq 1} |u_n(x,\mu) - u(x,\mu)| \to 0 \quad \text{as} \quad n \to \infty,$$

where

$$u_n(x,\mu) = \begin{cases} \int_a^x e^{\sigma/\mu(x-y)} \cdot \frac{\sigma}{\mu} G(\phi_n(y))dy, & \mu > 0; \\ G(\phi_n(x)), & \mu = 0; \\ -\int_{-a}^x e^{\sigma/\mu(x-y)} \cdot \frac{\sigma}{\mu} G(\phi_n(y))dy, & \mu < 0 \end{cases}$$

and sequence $\{\phi_n(x)\}$ is given by (3.3.24) with any initial, not identically zero function $\phi_0(x) \in D$.

Notice that the conclusions of Theorem 3.3.3 agree with physical facts. It is intuitively clear that for $c < 1$ no neutron has a nonzero probability of producing a persisting chain reaction. Moreover, since neutrons have a positive probability of leaving the reactor, the effective average number of neutrons emerging from a collision is decreased, and therefore it is only when $c > c^*$ for some $c^* > 1$ that the neutrons may have a positive probability of producing a persisting chain reaction.

The next theorem is concerned with the dependence of c^* on the width $2a$ of the reactor.

Theorem 3.3.4. The following three conclusions hold:

(a) c^* is a strictly decreasing and continuous function of a in $0 < a < +\infty$, and so $0 < a_1 < a_2$ implies $c_1^* > c_2^*$;

(b) $\lim_{a \to +0} c^* = +\infty$;

(c) $\lim_{a \to +\infty} c^* = 1$.

Proof. (a) Fix $p > 2$ and let $q = p/(p-1)$. By Holder's inequality, we have

$$0 < E(|x-y|) \leq \frac{\sigma}{2}\left(\int_{\sigma}^{+\infty} e^{-pt|x-y|}dt\right)^{1/p} \cdot \left(\int_{\sigma}^{+\infty} \frac{dt}{t^q}\right)^{1/q}$$

$$= \frac{\gamma e^{-\sigma|x-y|}}{|x-y|^{1/p}} \leq \frac{\gamma}{|x-y|^{1/p}}, \quad \forall\ x \neq y, \tag{3.3.31}$$

where

$$\gamma = 2^{-1} p^{-(1/p)} (q-1)^{-(1/q)} \sigma^{(1/q)} = \text{const.},$$

and therefore operator B (see (3.3.10)) is completely continuous from $L^2[-a,a]$ to $C[-a,a]$ (see Guo [4]). Consequently,

$$\frac{1}{c^*} = \max_{\|\phi\|_{L^2}=1} (B\phi,\phi), \tag{3.3.32}$$

where the maximum is attained at $\phi = \phi^* = \psi^*/\|\psi^*\|_{L^2}$. Now let $0 < a_1 < a_2$ be given. Then by virtue of (3.3.32) and Lemma 3.3.1 we see

$$\frac{1}{c_1^*} = (B\phi_1^*,\phi_1^*) = \int_{-a_1}^{a_1}\int_{-a_1}^{a_1} E(|x-y|)\phi_1^*(x)\phi_1^*(y)dxdy \ , \tag{3.3.33}$$

where $\phi_1^* = \psi^*/\|\psi_1^*\|_{L^2}$, $\phi_1^*(x) > 0$, $\phi_1^*(x) \in C[-a_1,a_1]$, $\|\phi_1^*\|_{L^2} = 1$ and

$$\phi_1^*(x) = c_1^* \int_{-a_1}^{a_1} E(|x-y|)\phi_1^*(y)dy, \quad -a_1 \leq x \leq a_1. \tag{3.3.34}$$

Define a function ϕ_0^* in $[-a_2,a_2]$ as follows:

$$\phi_0^*(x) = \begin{cases} \phi_1^*(x), & x \in [-a_1,a_1]; \\ 0\ , & x \in [-a_2,a_2] \setminus [-a_1,a_1]. \end{cases}$$

Then

$$\|\phi_2^*\|_{L^2} = \left(\int_{-a_2}^{a_2} |\phi_0^*(x)|^2 dx \right)^{1/2} = \left(\int_{-a_1}^{a_1} |\phi_1^*(x)|^2 dx \right)^{1/2} = 1.$$

It follows therefore from (3.3.32) and (3.3.33) that

$$\begin{aligned} \frac{1}{c^*} &= \max_{\|\phi\|_{L^2}=1} \int_{-a_2}^{a_2} \int_{-a_2}^{a_2} E(|x - y|)\phi(x)\phi(y)dxdy \\ &\geq \int_{-a_2}^{a_2} \int_{-a_2}^{a_2} E(|x - y|)\phi_0^*(x)\phi_0^*(y)dxdy \\ &= \int_{-a_1}^{a_1} \int_{-a_1}^{a_1} E(|x - y|)\phi_1^*(x)\phi_1^*(y)dxdy = \frac{1}{c_1^*}\ , \end{aligned}$$

and hence $c_1^* \geq c_2^*$. If $c_1^* = c_2^*$, then, by virtue of

$$\phi_2^*(x) = c_2^* \int_{-a_2}^{a_2} E(|x - y|)\phi_2^*(y)dy\ , \tag{3.3.35}$$

where $\phi_2^* = \psi^*/\|\psi_2^*\|_{L^2}$, $\phi_2^*(x) > 0$, $\phi_2^*(x) \in C[-a_2,a_2]$, $\|\phi_2^*\|_{L^2} = 1$, we obtain

$$\phi_2^*(x) \geq c_2^* E(a_2) \int_{-a_2}^{a_2} \phi_2^*(y)dy = \alpha_2 > 0, \quad -a_2 \leq x \leq a_2, \tag{3.3.36}$$

which implies that for sufficiently small $t > 0$,

$$t\phi_1^*(x) \leq \phi_2^*(x), \quad -a_1 \leq x \leq a_1.$$

Now, let $t_0 = \sup\{t > 0 \mid t\phi_1^*(x) \leq \phi_2^*(x),\ \forall\ -a_1 \leq x \leq a_1\}$. Then it is evident that $0 < t_0 < +\infty$ and $t_0\phi_1^*(x) \leq \phi_2^*(x)$, and therefore, by (3.3.34) and (3.3.35),

$$t_0\phi_1^*(x) \le c_1^* \int_{-a_1}^{a_1} E(|x - y|)\phi_2^*(y)dy = c_2^* \int_{-a_1}^{a_1} E(|x - y|)\phi_2^*(y)dy$$

$$< c_2^* \int_{-a_2}^{a_2} E(|x - y|)\phi_2^*(y)dy = \phi_2^*(x), \quad -a_1 \le x \le a_1.$$

Hence, for sufficiently small $\varepsilon > 0$, we have $(t_0 + \varepsilon)\phi_1^*(x) \le \phi_1^*(x)$ for $x \in [-a_1, a_1]$, which contradicts the definition of t_0. Thus, $c_1^* > c_2^*$.

To prove the continuity of c^*, let $0 < \ell_1 \le a_1 < a_2 \le \ell_2$, where ℓ_1 and ℓ_2 are fixed. Denoting $\int_{a_1}^{a_2} [\phi_2^*(x)]^2 dx = \delta/2$ and observing that $\phi_2^*(-x) = \phi_2^*(x)$, we find

$$\frac{1}{c_2^*} = \int_{-a_2}^{a_2} \int_{-a_2}^{a_2} E(|x - y|)\phi_2^*(x)\phi_2^*(y)dxdy$$

$$\le \int_{-a_1}^{a_1} \int_{-a_1}^{a_1} E(|x - y|)\phi_2^*(x)\phi_2^*(y)dxdy$$

$$+ \left(\int_{a_1}^{a_2} + \int_{-a_2}^{-a_1} \right)\phi_2^*(x)dx \int_{-a_2}^{a_2} E(|x - y|)\phi_2^*(x)dx$$

$$+ \left(\int_{a_1}^{a_2} + \int_{-a_2}^{-a_1} \phi_2^*(y)dy \int_{-a_2}^{a_2} E(|x - y|)\phi_2^*(x)dx \right.$$

$$= \int_{-a_1}^{a_1} \int_{-a_1}^{a_1} E(|x - y|)\phi_2^*(x)\phi_2^*(y)dxdy + \frac{4}{c_2^*} \int_{a_1}^{a_2} [\phi^*(x)]^2 dx$$

$$\le \frac{1 - \delta}{c_1^*} + \frac{2\delta}{c_2^*},$$

and so

$$0 < c_1^* - c_2^* \le \frac{\delta c_2^*}{1 - 2\delta}. \tag{3.3.37}$$

On the other hand, we get by (3.3.31)

$$\frac{\delta}{2} = \int_{a_1}^{a_2} \phi_2^*(x)\ {}^2 dx = (c_2^*)^2 \int_{a_1}^{a_2} dx \left[\int_{-a_2}^{a_2} E(|x - y|)\phi_2^*(y)dy \right]^2$$

$$\leq (c_2^*)^2 \int_{a_1}^{a_2} dx \int_{-a_2}^{a_2} E(|x - y|)^2 dy$$

$$\leq \gamma^2 (c_2^*)^2 \int_{a_1}^{a_2} dx \int_{-a_2}^{a_2} |x - y|^{-(2/p)} dy$$

$$= \frac{p\gamma^2 (c_2^*)^2}{p - 2} \int_{a_1}^{a_2} \left[(a_2 + x)^{1-(2/p)} + (a_2 - x)^{1-(2/p)} \right] dx$$

$$= \frac{pq\gamma^2 (c_2^*)^2}{2(p - 2)} \cdot \left[(2a_2)^{2/q} - (a_2 + a_1)^{2/q} + (a_2 - a_1)^{2/q} \right] . \tag{3.3.38}$$

In addition, by virtue of (3.3.36), we see that

$$\int_{-a_2}^{a_2} \phi_2^*(x)dx \geq 2a_2 c_2^* E(2a_2) \int_{-a_2}^{a_2} \phi_2^*(y)dy,$$

and so

$$1 \geq 2a_2 c_2^* E(2a_2) \geq 2\ell_1 c_2^* E(2\ell_2). \tag{3.3.39}$$

Now, it follows from (3.3.37), (3.3.38), and (3.3.39) that

$$0 < c_1^* - c_2^* \leq \frac{\delta'}{4\ell_1 E(2\ell_2)(1 - \delta')} ,$$

where

$$\delta' = \frac{pq\gamma^2}{2(p - 2)\ell_1^2\ E(2\ell_2)\ ^2} \cdot \left[(2a_2)^{2/q} - (a_2 + a_1)^{2/q} + (a_2 - a_1)^{2/q} \right],$$

which shows that c^* is a continuous function of a.

(b) and (c). First we show that

$$1 - \frac{1}{4\sigma a} + \frac{1}{4\sigma a} e^{-2\sigma a} - \frac{1}{2} e^{-2\sigma a} + 2aE(2a) \leq \frac{1}{c^*} \leq 1 - e^{-\sigma a} + 2aE(a). \tag{3.3.40}$$

From Lemma 3.3.1 (i), we find

$$\frac{1}{c^*} = \frac{\|B\psi^*\|}{\|\psi^*\|} \leq \|B\| = 1 - e^{-\sigma a} + 2aE(a),$$

it is the right-hand inequality of (3.3.40). On the other hand, letting $\phi_0(x) \equiv 1/\sqrt{2a}$ for $-a \leq x \leq a$, we see $\|\phi_0\|_{L^2} = 1$, and so by (3.3.32)

$$\frac{1}{c^*} \geq (B\phi_0,\phi_0) = 1 - \frac{\sigma}{a}\int_\sigma^{+\infty} \frac{e^{-ta}\sin hta}{t^3}\,dt. \tag{3.3.41}$$

Integrating by parts, it is easy to get

$$\int_\sigma^{+\infty} \frac{e^{-ta}\sin hta}{t^3}\,dt = \frac{1}{4\sigma^2} - \frac{1}{4\sigma^2}e^{-2\sigma a} + \frac{a}{2\sigma}e^{-2\sigma a} - \frac{ea^2}{\sigma}E(2a). \tag{3.3.42}$$

Now, the left-hand inequality of (3.3.40) follows from (3.3.41) and (3.3.42).

By Holder's inequality we obtain

$$aE(a) = \frac{\sigma a}{2}\int_\sigma^{+\infty}\frac{e^{-ta}}{t}\,dt \leq \frac{\sigma a}{2}\left(\int_\sigma^{+\infty} e^{-2ta}\,dt\right)^{1/2}\cdot\left(\int_\sigma^{+\infty}\frac{dt}{t^2}\right)^{1/2}$$

$$= \frac{\sqrt{\sigma a}}{2\sqrt{2}}e^{-\sigma a}.$$

Hence

$$\lim_{a\to+\infty} aE(a) = 0 \tag{3.3.43}$$

and

$$\lim_{a\to+0} aE(a) = 0\,. \tag{3.3.44}$$

Finally, the conclusion (b) follows from (3.3.44) and the right-hand inequality of (3.3.40), and conclusion (c) follows from (3.3.43) and (3.3.40).

Notice that the consequences of Theorem 3.3.4 agree with physical facts. For example, from $c = \sum_{k=1}^{N} kc_k \leq \sum_{k=1}^{N} k = \text{const}$ we see that c is bounded, and so the conclusion (b) of Theorem 3.3.4 shows that it cannot produce a persisting chain reaction when the reactor is too narrow. This is true in physics, since from the atomic physics we know that persisting chain reaction may be produced only when the volume of the reactor is greater than the so-called critical volume.

(b) A Nonlinear Integral Equation Arising in Nuclear Physics

Nonlinear integral equation

$$1 = \psi(x) + \psi(x) \int_0^1 \frac{R(x,y)}{x^2 - y^2} \psi(y)dy, \quad 0 \le x \le 1 \tag{3.3.45}$$

is of interest in nuclear physics (see Stuart [1]). The solution $\psi(x)$ of (3.3.45) satisfying $0 < \psi(x) \le 1$ can be interpreted as associating a probability $\psi(x)$ with the point $x \in [0,1]$. The formulation of problems in nuclear physics often takes this form.

Theorem 3.3.5. Suppose that (i) $R(x,y)$ is continuous on $0 \le x, y \le 1$ and $R(x,y) \ge 0$ for $x > y$ and $R(x,y) \le 0$ for $x < y$; (ii) there exists a $\nu > 0$ such that

$$|R(x,y)| \le c|x - y|^{\nu} S(x,y), \quad 0 \le x, \quad y \le 1, \quad x \ne y, \tag{3.3.46}$$

where c is a constant and $S(x,y)$ is a bounded and nonnegative function on $0 \le x, \; y \le 1$, which satisfies

$$\overline{\lim_{x,y\to+0}} \frac{S(x,y)}{x + y} < +\infty . \tag{3.3.47}$$

Then, equation (3.3.45) has exactly one positive continuous solution $\psi^*(x)$. Moreover, $0 < \psi^*(x) \le 1$ for $x \in [0,1]$ and, constructing successively the sequence of functions

$$\psi_{n+1}(x) = \left[1 + \int_0^1 \frac{R(x,y)}{x^2 - y^2} \psi_n(y)dy\right]^{-1} \quad (n = 0,1,2,\ldots) \tag{3.3.48}$$

for any initial function $\psi_0(x) \in C[0,1]$, which satisfies $0 < \psi_0(x) \le 1$, we have

$$\|\psi_n - \psi^*\| = \max_{0 \le x \le 1} |\psi_n(x) - \psi^*(x)| \to 0 \quad \text{as} \quad n \to \infty. \tag{3.3.49}$$

Proof. Since the theorem is obviously true for $R(x,y) \equiv 0$, we suppose that $R(x,y) \not\equiv 0$. Letting

$$\phi(x) = [\psi(x)]^{-1} - 1, \tag{3.3.50}$$

equation (3.3.45) reduces to

$$\phi(x) = \int_0^1 \frac{R(x,y)}{x^2 - y^2} \cdot \frac{1}{1 + \phi(y)}\, dy. \tag{3.3.51}$$

It is clear that $0 < \psi(x) \le 1$ is equivalent to $\phi(x) \ge 0$. Hence our problem reduces to the discussion of nonnegative continuous solution of Hammerstein integral equation (3.3.51).

By condition (ii) we see

$$\left|\frac{R(x,y)}{x^2 - y^2}\right| \quad \frac{c_1}{|x - y|^{1-\nu}}, \quad x \ne y, \quad 0 \le x, \quad y \le 1,$$

where c_1 is a constant and hence the linear integral operator

$$B\phi(x) = \int_0^1 \frac{R(x,y)}{x^2 - y^2}\, \phi(y)dy$$

is completely continuous from $C[0,1]$ into $C[0,1]$ (see Guo [4]), and so the nonlinear integral operator

$$A\phi(x) = \int_0^1 \frac{R(x,y)}{x^2 - y^2} \cdot \frac{1}{1 + \phi(y)}\, dy \tag{3.3.52}$$

maps the normal cone $P = \{\phi \in C[0,1] \mid \phi(x) \ge 0\}$ of $E = C[0,1]$ into P, and is completely continuous. Obviously, A is also decreasing, and therefore condition (i) of Theorem 2.1.5 is satisfied. Since $R(x,y) \not\equiv 0$, we have $A\theta > \theta$. Moreover, it is clear that $A^2\theta \ge \varepsilon_0 A\theta$, where

$$\varepsilon_0 = \frac{1}{1 + M}, \quad M = \|A\theta\| = \max_{0 \le x \le 1} \int_0^1 \frac{R(x,y)}{x^2 - y^2}\, dy\ .$$

Therefore, condition (ii) of Theorem 2.1.5 is satisfied. Now, let $\phi > \theta$ and $0 < t < 1$. We have

$$A[t\phi(x)] = \frac{1}{t}\int_0^1 \frac{R(x,y)}{x^2 - y^2} \cdot \frac{1}{t^{-1} + \phi(y)}\, dy. \tag{3.3.53}$$

Since ϕ is continuous and $t^{-1} + \phi(y) > 1 + \phi(y)$ for $0 \leq y \leq 1$, we see that

$$\min_{0 \leq y \leq 1} \frac{t^{-1} + \phi(y)}{1 + \phi(y)} = \; + \eta, \quad \eta > 0. \tag{3.3.54}$$

It follows from (3.3.53) and (3.3.54) that

$$A[t\phi(x)] \leq \frac{1}{t(1+\eta)}\int_0^1 \frac{R(x,y)}{x^2 - y^2} \cdot \frac{1}{1 + \phi(y)}\, dy = [t(1+\eta)]^{-1} A\phi(x),$$

i.e., condition (iii) of Theorem 2.1.5 is satisfied. Thus, by Theorem 2.1.5 A has exactly one positive fixed point ϕ^* and, constructing successively the sequence

$$\phi_{n+1} = A\phi_n \quad (n = 0,1,2,\ldots). \tag{3.3.55}$$

for any initial $\phi_0 \in P$, we have

$$\|\phi_n - \phi^*\| \to 0 \quad (n \to \infty). \tag{3.3.56}$$

Now, letting

$$\psi^*(x) = [1 + \phi^*(x)]^{-1}, \quad \psi_n(x) = [1 + \phi_n(x)]^{-1} \quad (n = 0,1,2,\ldots), \tag{3.3.57}$$

it is clear that $\psi^*(x)$ is the unique continuous solution of (3.3.45) satisfying $0 < \psi(x) \leq 1$ and, by (3.3.56), (3.3.49) holds. Finally, (3.3.48) follows directly from (3.3.55) and (3.3.57).□

Notice that it is easy to construct some elementary functions $R(x,y)$, which satisfy all conditions of Theorem 3.3.5. For example,

$$R(x,y) = C(x - y)^{1/3}\ (x + 3y - \sin x + 2x^2y),$$

and

$$R(x,y) = C(x - y)^{1/5}\ \ln\ (1 + x + y).$$

(c) A Nonlinear Integral Equation Modeling Infectious Disease. The nonlinear integral equation

$$x(t) = \int_{t-\tau}^{t} f(s,x(s))ds \tag{3.3.58}$$

can be interpreted as a model for the spread of certain infectious diseases with a periodic contact rate that varies seasonally (see Leggett and Williams [2] and Williams and Leggett [2]). In equation (3.3.58), $x(t)$ represents the proportion of infected individuals in the population at time t. $f(t,x(t))$ is the proportion of new infected individuals per unit time $(f(t,0) = 0)$, and τ is the length of time an individual remains infectious. Obviously, $x(t) \equiv 0$ is the trivial solution of equation (3.3.58), and we want to find another nontrivial periodic nonnegative solution.

Now, we formulate some conditions for the function $f(t,x)$:

(H_1) $f(t,x)$ is nonnegative and continuous for $-\infty < t < +\infty$ and $x \geq 0$;

(H_2) $f(t,0) = 0$ for $-\infty < t < +\infty$ and there exists $\omega \geq 0$ such that $f(t + \omega,x) = f(t,x)$ for all $-\infty < t < +\infty$ and $x \geq 0$;

(H_3) there exists $R > 0$ such that $f(t,x) \leq R/\tau$ for $0 \leq t \leq \omega$ and $0 \leq x \leq R$;

(H_4) $$\lim_{x \to +0} \frac{f(t,x)}{x} = a(t) \tag{3.3.59}$$

exists for all $t \in (-\infty,+\infty)$, and for each $0 < k < 1$ there exists $\varepsilon_k > 0$ such that

$$f(t,x) \geq ka(t)x, \quad -\infty < t < +\infty, \quad 0 \leq x \leq \varepsilon_k; \tag{3.3.60}$$

(H_5) $\overline{\lim\limits_{x\to+0}} \dfrac{f(t,x)}{x} = a_0(t)$ uniformly with respect to $t \in [0,\omega]$ and $\sup\limits_{0\leq t\leq\omega} a_0(t) < \dfrac{1}{\tau}$;

(H_6) $\overline{\lim\limits_{x\to+\infty}} \dfrac{f(t,x)}{x} = a_\infty(t)$ uniformly with respect to $t \in [0,\omega]$ and $\sup\limits_{0\leq t\leq\omega} a_\infty(t) < \dfrac{1}{\tau}$.

Theorem 3.3.6. Suppose that conditions $(H_1)(H_4)$ are satisfied and N is a positive integer such that $\omega/N < \tau/2$. Set $I_j = [\omega(j-1/N, \omega j/N]$ $(j = 0,1,2,\ldots,N)$. If

$$\prod_{j=1}^{N} \int_{I_j} a(s)ds > 1, \tag{3.3.61}$$

then equation (3.3.58) has at least one nontrivial nonnegative and continuous solution with period ω.

Proof. We first notice that by conditions (H_1)-(H_4), $a(t)$ is a nonnegative bounded measurable function with period ω, and so the integrals in (3.3.61) exist.

Let E be the Banach space of all continuous and ω-periodic functions $x(t)$ on R^1 with norm

$$\|x\| = \sup_{-\infty<t<+\infty} |x(t)| = \max_{0\leq t\leq\omega} |x(t)|,$$

and let

$$P = \{x(t) \in E \mid x(t) \geq 0 \ \text{ for } -\infty < t < +\infty\}.$$

Obviously, P is a cone of E. It is easy to show that the nonlinear integral operator

$$Ax(t) = \int_{t-\tau}^{t} f(s,x(s))ds \tag{3.3.62}$$

is completely continuous from P into P. Now, let $T_R = \{x(t) \in E \mid \|x\| < R\}$ and we will verify

$$Ax \not> (1 + \varepsilon)x, \ \forall \ x \in P \cap \partial T_R, \quad \varepsilon > 0. \tag{3.3.63}$$

In fact, if there exist $x_0 \in P \cap \partial T_R$ and $\varepsilon_0 > 0$ such that $Ax_0 \geq (1 + \varepsilon_0)x_0$, i.e., $Ax_0(t) \geq (1 + \varepsilon_0)x_0(t)$ for all $t \in R^1$, then, by (H_3), we have

$$(1 + \varepsilon_0)x_0(t) \leq \int_{t-\tau}^{t} f(s,x_0(s))ds \leq \int_{t-\tau}^{t} \frac{R}{\tau}\, ds = R, \ \forall \ t \in R^1,$$

and so $(1 + \varepsilon_0)\|x_0\| \leq R$, i.e., $(1 + \varepsilon_0)R \leq R$, which is impossible. Hence, (3.3.63) holds.

On the other hand, setting $u_0(t) \equiv 1$, we find $P_{u_0} = \{x(t) \in E \mid x(t) > 0 \text{ for } -\infty < t < +\infty\}$. From (3.3.61) we can choose $0 < k < 1$ such that

$$k^N \prod_{j=1}^{N} \int_{I_j} a(s)ds > 1, \tag{3.3.64}$$

and then choose $0 < r < R$ such that

$$f(t,x) \geq ka(t)x, \ \forall \ -\infty < t < +\infty, \quad 0 \leq x \leq r\ . \tag{3.3.65}$$

Now, we let $T_r = \{x(t) \in E \mid \|x\| < r\}$ and prove

$$Ax \not\leq x, \ \forall \ x \in P_{u_0} \cap \partial T_r. \tag{3.3.66}$$

Indeed, if there exists $x_1 \in P_{u_0} \cap \partial T_r$ such that $Ax_1 \leq x_1$, i.e., $Ax_1(t) \leq x_1(t)$ for all $t \in R^1$, then, observing $\omega/N \leq \tau/2$ so that $t \in I_j$ implies $I_{j-1} \subset [t - \tau,t]$, we get from (3.3.65)

$$\begin{aligned}\int_{I_j} a(t)x_1(t)dt &\geq \int_{I_j} a(t)Ax_1(t)dt = \int_{I_j} a(t)dt \int_{t-\tau}^{t} f(s,x_1(s))ds \\ &\geq \int_{I_j} a(t)dt \int_{I_{j-1}} f(s,x_1(s))ds \\ &\geq k\left(\int_{I_j} a(t)dt\right)\cdot\left(\int_{I_{j-1}} a(s)x_1(s)ds\right), \quad j = 1,2,\ldots,N,\end{aligned}$$

and therefore

$$\int_{I_N} a(t)x_1(t)dt \geq k^N\left(\prod_{j=1}^{N}\int_{I_j} a(t)dt\right)\cdot\left(\int_{I_0} a(t)x_1(t)dt\right)$$

$$= k^N\left(\prod_{j=1}^{N}\int_{I_j} a(t)dt\right)\cdot\left(\int_{I_N} a(t)x_1(t)dt\right). \qquad (3.3.67)$$

If $\int_{I_N} a(t)x_1(t)dt = 0$, then, on account of $a(t) \geq 0$ and $x_1(t) > 0$, we have $a(t) = 0$, p.p. in I_N, and so $\int_{I_N} a(t)dt = 0$, which contradicts (3.3.61). Hence, $\int_{I_N} a(t)x_1(t)dt > 0$, and therefore by (3.3.67)

$$k^N\left(\prod_{j=1}^{N}\int_{I_j} a(t)dt\right) \leq 1,$$

in contradiction with (3.3.64). Thus, (3.3.66) holds.

Finally, by Theorem 2.3.5 we know that A has at least one fixed point in $P \cap (\overline{T}_R \setminus T_r)$ and our theorem is proved. □

Theorem 3.3.7. Suppose that (H_1), (H_2), (H_5), and (H_6) are satisfied. If there exist $a > 0$ and a nonnegative continuous function $b(t)$ with period ω (i.e., $b(t + \omega) = b(t)$ for $-\infty < t < +\infty$) such that

$$f(t,x) \geq b(t), \quad t \in [0,\omega], \quad x \geq a \qquad (3.3.68)$$

and

$$\min_{0\leq t\leq\omega} \int_{t-\tau}^{t} b(s)ds > a, \qquad (3.3.69)$$

then equation (3.3.58) has at least two nontrivial nonnegative and continuous solutions $x_1(t)$ and $x_2(t)$ with period ω and

$$\inf_{-\infty<t<+\infty} x_1(t) > a\ . \qquad (3.3.70)$$

Proof. As in the proof of Theorem 3.3.6, we can show easily that the operator A (see (3.3.62)) is completely continuous from P into P. Choosing $\mu > \tau$ such that

$$\sup_{0 \le t \le \omega} a_0(t) < \frac{1}{\mu}, \quad \sup_{0 \le t \le \omega} a_\infty(t) < \frac{1}{\mu}$$

and observing that $a_0(t)$ and $a_\infty(t)$ are ω-periodic functions, we see by conditions (H_5) and (H_6) that there exist $0 < r < a$ and $\ell > a$ such that

$$0 \le f(t,x) < \frac{1}{\mu} x, \quad -\infty < t < +\infty, \quad 0 < x \le r \quad \text{or} \quad x \ge \ell. \tag{3.3.71}$$

Hence

$$0 \le f(t,x) \le \frac{1}{\mu} x + \beta, \quad -\infty < t < +\infty, \quad x \ge 0, \tag{3.3.72}$$

where

$$\beta = \max_{0 \le t \le \omega,\, 0 \le x \le \ell} f(t,x) .$$

Now, select $R > a$ such that $R/\mu + \beta \le R/\tau$. Then from (3.3.71) and (3.3.72) it follows that

$$0 \le f(t,x) < \frac{r}{\mu}, \quad -\infty < t < +\infty, \quad 0 \le x \le r \tag{3.3.73}$$

and

$$0 \le f(t,x) < \frac{R}{\tau}, \quad -\infty < t < +\infty, \quad 0 \le x \le R \ . \tag{3.3.74}$$

Let $U_1 = \{x \in P \mid \|x\| < r\}$, $U_2 = \{x \in P \mid \|x\| < R$ and $\min_{0 \le t \le \omega} x(t) > a\}$ and $U_3 = \{x \in P \mid \|x\| \le R\}$. Obviously,

$$U_1 \subset U_3, \quad U_2 \subset U_3, \quad U_1 \cap U_2 = \phi. \tag{3.3.75}$$

and $\overline{U}_1 = \{x \in P \mid \|x\| \le r\}$, $\overline{U}_2 = \{x \in P \mid \|x\| \le R$ and $\min_{0 \le t \le \omega} x(t) \ge a\}$, $\overline{U}_3 = \{x \in P \mid \|x\| \le R\}$. For $x \in \overline{U}_1$, from (3.3.73) we have

$$\|Ax\| < \max_{0\le t\le \omega} \int_{t-\tau}^{t} \frac{r}{\mu}\, ds = \frac{r\tau}{\mu} < r,$$

and hence

$$A(\overline{U}_1) \subset U_1. \tag{3.3.76}$$

By the same way, for $x \in \overline{U}_3$ we see from (3.3.74)

$$\|Ax\| < \max_{0\le t\le \omega} \int_{t-\tau}^{t} \frac{R}{\tau}\, ds = R,$$

and

$$A(\overline{U}_3) \subset U_3. \tag{3.3.77}$$

When $x \in \overline{U}_2$, we have $\|x\| \le R$ and $\inf\limits_{-\infty<t<+\infty} x(t) = \min\limits_{0\le t\le\omega} x(t) \ge a$ and, by (3.3.77), $\|Ax\| < R$. Moreover, by virtue of (3.3.68) and (3.3.69) we get

$$\min_{0\le t\le\omega} Ax(t) \ge \min_{0\le t\le\omega} \int_{t-\tau}^{t} b(s)ds > a,$$

which implies

$$A(\overline{U}_2) \subset U_2 . \tag{3.3.78}$$

It follows from (3.3.76), (3.3.77), (3.3.78), and Lemma 2.4.1 that

$$i(A, U_k, P) = 1, \quad k = 1,2,3 . \tag{3.3.79}$$

Thus, A has a fixed point x_1 in U_2. It is clear that

$$\inf_{-\infty<t<+\infty} x_1(t) = \min_{0\le t\le\omega} x_1(t) > a ,$$

i.e., (3.3.70) holds. On the other hand, from (3.3.79) we see that

$$i(A, U_3 \setminus (\overline{U}_1 \cup \overline{U}_2), P) = i(A, U_3, P) - i(A, U_1, P) - i(A, U_2, P) = -1 \ne 0,$$

and hence A has a fixed point x_2 in $U_3 \setminus (\overline{U}_1 \cup \overline{U}_2)$, and our theorem is proved.□

It is easy to give some elementary functions, which satisfy all conditions of Theorem 3.3.7. For example,

$$f(t,x) = b(t)\sqrt{x}\ \ln(1+x), \quad -\infty < t < +\infty, \quad u \geq 0 \tag{3.3.80}$$

where $b(t)$ is a nonnegative continuous function with period ω, which satisfies

$$\min_{0 \leq t \leq \omega} \int_{t-\tau}^{t} b(s)ds > \frac{e}{2} ;$$

and

$$f(t,x) = c(t)\left(\frac{x^2}{1+x^2} + \sin^2 x \right), \quad \forall \ -\infty < t < +\infty, \quad x \geq 0, \tag{3.3.81}$$

where $c(t)$ is a nonnegative continuous function with period ω, which satisfies

$$\min_{0 \leq t \leq \omega} \int_{t-\tau}^{t} c(s)ds > 2 .$$

For functions (3.3.80) and (3.3.81) we take $a = e^2 - 1$ and $a = 1$ in (3.3.68) and (3.3.69), respectively.

Theorem 3.3.8. Suppose that (H_1)-(H_3) are satisfied. If there exist $0 < a < R$ and a nonnegative continuous function $b(t)$ with period such that

$$f(t,x) \geq b(t), \quad \forall\ t \in [0,\omega], \quad a \leq x \leq R \tag{3.3.82}$$

and

$$\min_{0 \leq t \leq \omega} \int_{t-\tau}^{t} b(s)ds \geq a. \tag{3.3.83}$$

Then equation (3.3.58) has at least one positive and continuous solution $x(t)$ with period ω and

$$a \leq \inf_{-\infty<t<+\infty} x(t) \leq \sup_{-\infty<t<+\infty} x(t) \leq R \, . \tag{3.3.84}$$

Proof. Let A,E,P and U_2 be the same as in the proof of Theorem 3.3.7. Obviously, $\overline{U}_2 = \{x \in P \mid a \leq x(t) \leq R, \ \forall \ t \in [0,\omega]\}$ and $\overline{U}_2$ is a nonempty bounded closed convex set in E. For $x \in \overline{U}_2$, we see by (H_3) and (3.3.82), (3.3.83) that

$$Ax(t) \leq \int_{t-\tau}^{t} \frac{R}{\tau} \, ds = R, \ \forall \ t \in [0,\omega]$$

and

$$\min_{0 \leq t \leq \omega} Ax(t) \geq \min_{0 \leq t \leq \omega} \int_{t-\tau}^{t} b(s) ds \geq a, \ \forall \ t \in [0,\omega] \, ,$$

and hence $Ax \in \overline{U}_2$, and therefore A is a completely continuous operator from $\overline{U}_2$ into $\overline{U}_2$ and, by Schauder's fixed point theorem, A has a fixed point in $\overline{U}_2$. □

It is easy to give some elementary functions which satisfy all conditions of Theorem 3.3.8. For example,

$$f(t,x) = \sum_{i=1}^{n} b_i(t) x^{\alpha_i} \, , \tag{3.3.85}$$

where $0 < \alpha_i < 1$ for $\alpha = 1,2,\ldots,n$ and $b_i(t)$ $(i = 1,2,\ldots,n)$ are nonnegative continuous functions with period ω and

$$\min_{0 \leq t \leq \omega} \int_{t-\tau}^{t} \left[\sum_{i=1}^{n} b_i(s) \right] ds \geq 1 \; ;$$

and

$$f(t,x) = b(t) \ln(1 + x^5) + c(t)\sqrt{x} \, \sin^2\left(x + \frac{\pi}{\omega} t\right), \tag{3.3.86}$$

where $b(t)$ and $c(t)$ are nonnegative continuous functions with period ω and $b(t)$ satisfies

$$\min_{0\le t\le\omega} \int_{t-\tau}^{t} b(s)ds \ge (\ln 2)^{-1} .$$

Notice that conditions (3.3.82) and (3.3.83) of Theorem 3.3.8 are essentially different from conditions (H_4) and (3.3.61) of Theorem 3.3.6, because (3.3.82) is concerned with $f(t,x)$ for $a \le x \le R$, where $a > 0$, and (3.3.61) deals with the values of $f(t,x)$ for sufficiently small $x > 0$.

3.4 *Infinitely Many Solutions Obtained by Variational Methods.*

Consider the Hammerstein integral equation

$$\phi(x) = \int_G k(x,y)f(y,\phi(y))dy = A\phi(x), \tag{3.4.1}$$

where G is a measurable set in Euclidean space with $0 < \text{mes}\ G < +\infty$ and $f(x,u)$ satisfies the Caratheodory condition, i.e., $f(x,u)$ is measurable with respect to x on G for every $u \in (-\infty,+\infty)$ and continuous with respect to u on $(-\infty,+\infty)$ for almost all $x \in G$. Suppose that the kernel $k(x,y)$ is measurable on $G \times G$ and

$$\int_G \int_G k(x,y)|^p\ dxdy < +\infty \tag{3.4.2}$$

for some $p > 2$, then the linear integral operator

$$k\phi(x) = \int_G k(x,y)\phi(y)dy \tag{3.4.3}$$

is completely continuous from $L^q(G)$ into $L^p(G)$, where $1/p + 1/q = 1$, $1 < q < 2 < p$. The following conditions are needed:

(P_1) Symmetric kernel $k(x,y)$ satisfies (3.4.2) and is strictly positive-definite on $L^q(G)$, i.e.,

$$(K\phi,\phi) = \int_G \int_G k(x,y)\phi(x)\phi(y)\, dxdy > 0, \quad \forall\ \phi \in L^q(G), \quad \phi \neq \theta, \tag{3.4.4}$$

where θ denotes the zero element;

(Q_1) There exist $a > 0$ and $b > 0$ such that

$$|f(x,u)| \leq a + b|u|^{p-1}, \ \forall\ x \in G, \quad -\infty < u < +\infty. \tag{3.4.5}$$

(Q_2) There exist $0 \leq \tau < 1/2$ and $M > 0$ such that

$$F(x,y) = \int_0^u f(x,v)dv \leq \tau u f(x,u), \ \forall\ x \in G, \quad |u| \geq M; \tag{3.4.6}$$

(Q_3) $f(x,u)/u \to 0$ as $u \to 0$ uniformly in $x \in G$;

(Q_4) $f(x,u)/u \to +\infty$ as $u \to \infty$ uniformly in $x \in G$;

(Q_5) $f(x,u)$ is odd in u, i.e.,

$$f(x,-u) = -f(x,u), \ \forall\ x \in G, \quad -\infty < u < +\infty. \tag{3.4.7}$$

Ambrosetti and Rabinowitz have considered (see Ambrosetti and Rabinowitz [1], Theorems 5.8 and 5.9) the following situations:

(A) if (P_1), (Q_1)-(Q_4) are satisfied, then the integral equation (3.4.1) has at least one nontrivial solution in $L^p(G)$;

(B) if (P_1), (Q_1)-(Q_5) are satisfied, then equation (3.4.1) has infinite many nontrivial solutions in $L^p(G)$.

However, many important kernels do not satisfy condition (P_1), such as boundary value problems. Consequently, we shall weaken condition (P_1) to overcome this shortcoming and also weaken conditions (Q_3) and (Q_4). We shall deal with two cases of positive-definite kernels and quasi-positive-definite kernels and give some applications to the two-point boundary value problems of nonlinear ordinary differential equations.

(a) Case of Positive-Definite Kernels.

Definition 3.4.1. (i) $k(x,y)$ is said to be a L^2 kernel, if $k(x,y)$ is measurable on $G \times G$, $\text{mes}\{(x,y) \in G \times G \mid k(x,y) \neq 0\} > 0$ and

$$\int_G \int_G [k(x,y)]^2 \, dxdy < +\infty . \tag{3.4.8}$$

(ii) L^2 kernel $k(x,y)$ is said to be positive-definite (quasi-positive-definite), if $k(x,y)$ is symmetric and all its nonzero eigenvalues are positive (i.e., if it only has a finite number of negative eigenvalues).

(iii) L^2 kernel $k(x,y)$ is said to be strictly positive-definite if it is symmetric and satisfies

$$(K\psi,\psi) = \int_G \int_G k(x,y)\psi(x)\psi(y) \, dxdy > 0, \quad \psi \in L^2(G), \quad \psi \neq \theta. \tag{3.4.9}$$

It is well known (see Zaanen [1]) that L^2 symmetric kernel is positive-definite if and only if

$$(K\psi,\psi) = \int_G \int_G k(x,y)\psi(x)\psi(y) \, dxdy \geq 0, \quad \psi \in L^2(G),$$

i.e., K is a positive operator. Since $\text{mes}\, G < +\infty$ and $1 < q < 2 < p$ imply $L^2(G) \subset L^q(G)$, we obtain the following:

$$\begin{aligned} k(x,y) \text{ satisfies } (\mathrm{P}_1) &\Rightarrow k(x,y) \text{ is strictly positive-definite} \\ &\Rightarrow k(x,y) \text{ is positive-definite.} \end{aligned} \tag{3.4.10}$$

Lemma 3.4.1. Suppose that the kernel $k(x,y)$ is positive-definite. Denote the sequence of eigenvalues and the corresponding sequence of orthonormal eigenfunctions of $k(x,y)$ by $\{\lambda_n\}$ $(\lambda_1 \geq \lambda_2 \geq \ldots > 0)$ and $\{\lambda_n\}$, respectively. Let $H_0 = \{\psi \in L^2(G) \mid K^{1/2}\psi = \theta\}$ and $H_1 = H_0$, where $K^{1/2}$ denotes the positive square root operator of K and $H_0^{\perp}$ denotes the orthogonal complement in $L^2(G)$ of $H_0^{\perp}$. Then

(i) $H_0 = \{\psi \in L^2(G) \mid K\psi = \theta\}$;

(ii) $H_1 = \overline{L\{\psi_1,\psi_2,\ldots\}}$, i.e., H_1 is the closed subspace spanned by $\psi_1,\psi_2,\ldots$;

(iii) $H_1 \neq \{\theta\}$, and so K has at least one eigenvalue.

Proof. (i) follows immediately from the equality

$$(K\psi,\psi) = (K^{1/2}\psi,K^{1/2}\psi) = \|K^{1/2}\psi\|^2 .$$

To prove (ii), it is sufficient to show

$$\overline{L\{\psi_1,\psi_2,\ldots\}}^{\perp} = H_0 . \tag{3.4.11}$$

It is well known that

$$K\psi = \sum_n \lambda_n(\psi,\psi_n)\psi_n, \quad \forall \ \psi \in L^2(G) \tag{3.4.12}$$

and therefore, $\psi \in \overline{L\{\psi,\psi_2,\ldots\}}^{\perp}$ implies $K\psi = \theta$ and hence $\psi \in H_0$ by (i). Conversely, $\psi \in H_0$ implies

$$\theta = K\psi = \sum_n \lambda_n(\psi,\psi_n)\psi_n ,$$

and so

$$0 = (K\psi,\psi_m) = \lambda_m(\psi,\psi_m), \quad m = 1,2,3,\ldots .$$

Hence, on account of $\lambda_m > 0$, $\psi \in \overline{L\{\psi_1,\psi_2,\ldots\}}^{\perp}$.

Finally, if $H_1 = \{\theta\}$, then $H_0 = L^2(G)$. It follows from (i) that $K = \theta$ and therefore $k(x,y) = 0$ p.p. on $G \times G$, in contradiction to Definition 3.4.1. Hence, (iii) holds and the lemma is proved. □

It follows from (ii) of Lemma 3.4.1 that positive-definite kernel $k(x,y)$ is strictly positive-definite if and only if its sequence of orthonormal eigenfunctions $\{\psi_n\}$ is complete in $L^2(G)$.

In this paragraph, $\{\lambda_n\}$ $(\lambda_1 \geq \lambda_2 \geq \dots > 0)$ and $\{\psi_n\}$ always denote the sequence of eigenvalues and the corresponding sequence of orthonormal eigenfunctions of the positive-definite kernel $k(x,y)$ respectively. The following conditions are used:

(P_2) $k(x,y)$ is positive-definite and satisfies (3.4.2) for some $p > 2$;

(Q_3') there exist $\varepsilon_0 > 0$ and $\delta > 0$ such that

$$\frac{f(x,u)}{u} \leq \frac{1}{\lambda_1 + \varepsilon_0}, \quad x \in G, \quad 0 < |u| < \delta; \tag{3.4.13}$$

(Q_4') there exist $0 < \varepsilon_0 < \lambda_1$ and $R > 0$ such that

$$\frac{f(x,u)}{u} \geq \frac{1}{\lambda_1 - \varepsilon_0}, \quad x \in G, \quad |u| \geq R; \tag{3.4.14}$$

(Q_3'') $\dfrac{f(x,u)}{u} \to a_0(x)$ as $u \to 0$ uniformly in $x \in G$, where $a_0(x)$ satisfies

$$\sup_{x\in G} a_0(x) < \frac{1}{\lambda_1};$$

(Q_4'') $\dfrac{f(x,u)}{u} \to a_1(x)$ as $u \to \infty$ uniformly in $x \in G$, where $a_1(x)$ satisfies

$$\inf_{x\in G} a_1(x) > \frac{1}{\lambda_1}.$$

Obviously, $(P_1) \Rightarrow (P_2)$, $(Q_3) \Rightarrow (Q_3'') \Rightarrow (Q_3')$ and $(Q_4) \Rightarrow (Q_4'') \Rightarrow (Q_4')$.

<u>Theorem 3.4.1</u>. If conditions (P_2), (Q_1), (Q_2), (Q_3') and (Q_4') are satisfied, then the integral equation (3.4.1) has at least one nontrivial solution in $L^p(G)$.

<u>Proof</u>. We have $A = Kf$, where K is the linear integral operator (3.4.3) and f is the operator $f\phi(x) = f(x,\phi(x))$. From (P_2) and (Q_1) we know that K is completely continuous from $L^2(G)$ into $L^2(G)$ and from

$L^q(G)$ into $L^p(G)$ $(1/p + 1/q = 1)$ and f is bounded and continuous from $L^p(G)$ into $L^q(G)$ (see Krasnosel'skii [2] and Guo [17]); moreover, we have $K = HH^*$, where $H = K^{1/2}$ is completely continuous from $L^2(G)$ into $L^p(G)$ and H^* denotes the adjoint operator of H, which is completely continuous from $L^q(G)$ into $L^2(G)$ (see Vainberg [1]). It is well known (see Krasnosel'skii [2]) that the functional

$$\Psi(\psi) = \frac{1}{2}(\psi,\psi) - \int_G F(x,H\psi)dx, \quad \forall \ \psi \in L^2(G) \tag{3.4.15}$$

is a C^1 functional in $L^2(G)$ and its Frechet derivative is

$$\Psi'(\psi) = \psi - H^*fH\psi. \tag{3.4.16}$$

Let H_0 and H_1 be the closed subspaces of $L^2(G)$ mentioned in Lemma 3.4.1. By Lemma 3.4.1 we see $H_1 \neq \{\theta\}$ and $H_1 = H_0^{\perp}$. Since

$$(H^*fH\psi,h) = (fH\psi,Hh) = (fH\psi,K^{1/2}h) = 0, \quad \forall \ \psi \in L^2(G), \ h \in H_0,$$

it follows $H^*fH\psi \in H_1$ for all $\psi \in L^2(G)$ and hence, we can regard H^*fH as an operator from H_1 into H_1 and (3.4.16) still holds for all $\psi \in H_1$ when we regard Ψ as a functional only on H_1. Next, we verify that the functional Ψ (on H_1) satisfies all conditions of the mountain pass lemma, i.e., conditions (I_1), (I_2), and (I_3) of Theorem 2.1 in Ambrosetti and Rabinowitz [1]; this theorem is also true for finite-dimensional space, (see Nirenberg [2]).

First we verify (I_1). (Q_3') implies

$$F(x,u) \leq \frac{u^2}{2(\lambda_1 + \varepsilon_0)}, \quad \forall \ x \in G, \quad |u| < \delta \tag{3.4.17}$$

and (Q_1) implies

$$F(x,u) \leq a|u| + \frac{b}{p}|u|^p, \quad \forall \ x \in G, \quad -\infty < u < +\infty . \tag{3.4.18}$$

Consequently, there exists $b_1 > 0$ such that

$$F(x,u) \leq \frac{u^2}{2(\lambda_1 + \varepsilon_0)} + b_1|u|^p, \quad x \in G, \quad -\infty < u < +\infty.$$

Thus, for $\psi \in H_1$, we have

$$\begin{aligned}
\int_G F(x,H\psi)dx &\leq \frac{1}{2(\lambda_1 + \varepsilon_0)} \int_G [H\psi(x)]^2 dx + b_1 \int_G |H\psi(x)|^p \, dx \\
&= \frac{1}{2(\lambda_1 + \varepsilon_0)} \|K^{1/2}\psi\|^2 + b_1\|H\psi\|_p^p \\
&= \frac{1}{2(\lambda_1 + \varepsilon_0)} (K\psi,\psi) + b_1\|H\psi\|_p^p \\
&\leq \frac{}{2(\lambda_1 + \varepsilon_0)} \|K\| \cdot \|\psi\|^2 + b_1\|H\psi\|_p^p \\
&= \frac{\lambda_1}{2(\lambda_1 + \varepsilon_0)} \|\psi\|^2 + b_1\|H\psi\|_p^p \, ,
\end{aligned} \tag{3.4.20}$$

where $\|\cdot\|_p$ denotes the norm in $L^p(G)$ and $\|K\|$ denotes the norm of operator K, which satisfies $\|K\| = \lambda_1$ (see Zaanen [1]). From (3.4.15) and (3.4.20) we get

$$\Psi(\psi) \geq \frac{\varepsilon_0}{(\lambda_1 + \varepsilon_0)} \|\psi\|^2 - b_1\|H\|^p \cdot \|\psi\|^p, \quad \forall \quad \psi \in H_1. \tag{3.4.21}$$

and hence, on account of $p > 2$, there exists sufficiently small $r > 0$ such that

$$\begin{aligned}
&\Psi(\psi) > 0, \ \forall \ \psi \in B_r \setminus \{\theta\} \, , \\
&\inf_{\psi \in \partial B_r} \Psi(\psi) = c_r > 0 \, ,
\end{aligned} \tag{3.4.22}$$

where $B_r = \{\psi \in H_1 \mid \|\psi\| < r\}$, i.e., condition (I_1) is satisfied.

Second, we verify (I_2). Consider the continuous function

$$\begin{aligned}
\Phi(t) = \Psi(t\psi_1) &= \frac{t^2}{2} \|\psi_1\|^2 - \int_G F(x, tH\psi_1)dx \\
&= \frac{t^2}{2} - \int_G F(x, t\sqrt{\lambda_1} \; \psi_1)dx,
\end{aligned} \tag{3.4.23}$$

where we have used the known equality $H\psi_1 = K^{1/2}\psi_1 = \sqrt{\lambda_1}\ \psi_1$. Choosing $0 < t_1 < t_2 < \ldots,\ t_n \to +\infty$, and letting $D_n = \{x \in G \mid t_n\sqrt{\lambda_1}\ |\psi_1(x)| \geq R\}$, we see $G_1 D_n = \{x \in G \mid t_n\sqrt{\lambda_1}\ |\psi_1(x)| < R\}$ and $D_n = D_n^{(1)} \cup D_n^{(2)}$ with $D_n^{(1)} \cap D_n^{(2)} = \phi$, where $D_n^{(1)} = \{x \in G \mid t_n\sqrt{\lambda_1}\ \psi_1(x) \geq R\}$ and $D_n^{(2)} = \{x \in G \mid t_n\sqrt{\lambda_1}\ \psi_1(x) \leq -R\}$. By ($Q_4'$) we know

$$f(x,u) \geq \frac{u}{\lambda_1 - \varepsilon_0}, \quad \forall\ x \in G, \quad u \geq R,$$

and

$$f(x,u) \leq \frac{u}{\lambda_1 - \varepsilon_0}, \quad \forall\ x \in G, \quad u \leq -R,$$

and hence

$$\begin{aligned}
\int_G F(x,t_n\sqrt{\lambda_1}\ \psi_1)dx &= \int_{D_n^{(1)}} dx\left(\int_0^R + \int_R^{t_n\sqrt{\lambda_1}\psi_1(x)} f(x,v)dv\right) \\
&+ \int_{D_n^{(2)}} dx\left(\int_0^{-R} + \int_{-R}^{t_n\sqrt{\lambda_1}\psi_1(x)} f(x,v)dv\right) \\
&+ \int_{G\setminus D_n} dx \int_0^{t_n\sqrt{\lambda_1}\psi_1(x)} f(x,v)dv \\
&\geq \int_{D_n^{(1)}} dx \int_R^{t_n\sqrt{\lambda_1}\psi_1(x)} \frac{v}{\lambda_1 - \varepsilon_0}\, dv \\
&- \int_{D_n^{(1)}} dx \int_0^R |f(x,v)|dv \\
&- \int_{D_n^{(2)}} dx \int_{t_n\sqrt{\lambda_1}\psi_1(x)}^{-R} \frac{v}{\lambda_1 - \varepsilon_0}\, dv \\
&- \int_{D_n^{(2)}} dx \int_{-R}^0 |f(x,v)|dv \\
&- \int_{G\setminus D_n} dx \int_{-R}^R |f(x,v)|dv
\end{aligned}$$

$$\geq \frac{1}{2(\lambda_1 - \varepsilon_0)} \int_{D_n^{(1)}} \{t_n^2 \lambda_1 [\psi_1(x)]^2 - R^2\} dx$$

$$- \int_{D_n} dx \int_0^R |f(x,v)| dv$$

$$+ \frac{1}{2(\lambda_1 - \varepsilon_0)} \int_{D_n^{(2)}} \{t_n^2 \lambda_1 [\psi_1(x)]^2 - R^2\} dx$$

$$- \int_{D_n} dx \int_{-R}^0 |f(x,v)| dv$$

$$- \int_{G \setminus D_n} dx \int_{-R}^R |f(x,v)| dv$$

$$= \frac{1}{2(\lambda_1 - \varepsilon_0)} \int_{D_n} \{t_n^2 \lambda_1 [\psi_1(x)]^2 - R^2\} dx$$

$$- \int_G dx \int_{-R}^R |f(x,v)| dv$$

$$\geq \frac{\lambda_1 t_n^2}{2(\lambda_1 - \varepsilon_0)} \int_{D_n} [\psi_1(x)]^2 dx - M_1 ,$$

where

$$M_1 = \frac{R^2}{2(\lambda_1 - \varepsilon_0)} + 2aR + \frac{2b}{p} R^p \cdot \text{mes } G = \text{const.}$$

Setting $D = \{x \in G \mid \psi_1(x) \neq 0\}$, we find

$$\int_D [\psi_1(x)]^2 dx = \int_G [\psi_1(x)]^2 dx = \|\psi_1\|^2 = 1$$

and $D_1 \subset D_2 \subset D_3 \subset \cdots$, $\bigcup_{n=1}^{\infty} D_n = D$, $\text{mes } D_n \to \text{mes } D$ $(n \to \infty)$, and therefore there exists $N_0 > 0$ such that

$$\int_{D_n} [\psi_1(x)]^2 dx > \int_D [\psi_1(x)]^2 dx - \frac{\varepsilon_0}{2\lambda_1} = 1 - \frac{\varepsilon_0}{2\lambda_1} , \ \forall \ n > N_0. \tag{3.4.25}$$

It follows from (3.4.23), (3.4.24), and (3.4.25) that

$$\Phi(t_n) = \frac{t_n^2}{2} - \int_G F(x,t_n\sqrt{\lambda_1}\,\psi_1)dx < -\frac{\varepsilon_0}{4(\lambda_1 - \varepsilon_0)}\,t_n^2 + M_1, \;\forall\; n > N_0, \tag{3.4.26}$$

which implies $\Phi(t_n) \to -\infty$ as $n \to \infty$. On the other hand, (3.4.22) implies $\Phi(r) = \Psi(r\psi_1) \geq c_r > 0$. It follows therefore from the continuity of $\Phi(t)$ that there exists $r < t^* < +\infty$ such that $\Phi(t^*) = \Psi(t^*\psi_1) = 0$ and, thus, (I_2) is satisfied by taking $e = t^*\psi_1$.

Finally we verify (I_3). In fact, we can prove that Ψ satisfies the (P.S.) condition. Let $\{h_n\} \subset H_1$ with $|\Psi(h_n)| \leq \beta$ $(n = 1,2,3,\ldots)$ and $\Psi'(h_n) \to \theta$. Setting $G_n = \{x \in G \mid |Hh_n(x)| \geq M\}$, we find from (Q_1) and (Q_2) that

$$\begin{aligned}\beta \geq \Psi(h_n) &= \frac{1}{2}\|h_n\|^2 - \int_G F(x,Hh_n)dx \\ &\geq \frac{1}{2}\|h_n\|^2 - \int_{G_n} F(x,Hh_n)dx - M_2 \\ &\geq \frac{1}{2}\|h_n\|^2 - \tau\int_{G_n} f(x,Hh_n)Hh_n dx - M_2 \\ &\geq \frac{1}{2}\|h_n\|^2 - \tau\int_G f(x,Hh_n)Hh_n dx - M_3,\end{aligned} \tag{3.4.27}$$

where M_2 and M_3 are constants independent of n. By virtue of (3.4.16), we have

$$\begin{aligned}(\Psi'(h_n),h_n) &= (h_n - H^*fHh_n,h_n) = (h_n,h_n) - (fHh_n,Hh_n) \\ &= \|h_n\|^2 - \int_G f(x,Hh_n)Hh_n dx.\end{aligned} \tag{3.4.28}$$

It follows from (3.4.27) and (3.4.28) that

$$\begin{aligned}\beta &\leq \left(\frac{1}{2} - \tau\right)\|h_n\|^2 + \tau(\Psi'(h_n),h_n) - M_3 \\ &\geq \left(\frac{1}{2} - \tau\right)\|h_n\|^2 - \tau\|\Psi'(h_n)\|\cdot\|h_n\| - M_3 \quad (n = 1,2,3,\ldots),\end{aligned} \tag{3.4.29}$$

and hence $\{h_n\}$ is bounded. Since H_1 is a Hilbert space, there exists a subsequence $\{h_{n_k}\} \subset \{h_n\}$, that converges weakly to an element $h_0 \in H_1$, and so, on account of the complete continuity of H, $\|Hh_{n_k} - Hh_0\|_p \to 0$. It therefore implies from (3.4.28) and the continuity of operator f that

$$\|h_{n_k}\|^2 \to \int_G f(x,Hh_0)Hh_0dx = (fHh_0,Hh_0). \tag{3.4.30}$$

On the other hand, we have

$$(\Psi'(h_{n_k}),h_0) = (h_{n_k} - H^*fHh_{n_k},h_0) = (h_{n_k},h_0) - (fHh_{n_k},Hh_0).$$

Taking the limit as $k \to \infty$, we get

$$\|h_0\|^2 = (fHh_0,Hh_0) . \tag{3.4.31}$$

We get from (3.4.30) and (3.4.31) that $\|h_{n_k}\| \to \|h_0\|$ as $k \to \infty$, and hence $\|h_{n_k} - h_0\| \to 0$ as $k \to \infty$, and (I_3) is satisfied.

Now, by Theorem 2.1 in Ambrosetti and Rabinowitz [1], Ψ has a critical point $\psi^* \in H_1$ with $\psi^* \neq \theta$, i.e.,

$$\Psi'(\psi^*) = \psi^* - H^*fH\psi^* = \theta.$$

Letting $\phi^* = H\psi^* \in L^p(G)$, we find

$$A\phi^* = Kf\phi^* = HH^*fH\psi^* = H\psi^* = \phi^* ,$$

i.e., ϕ^* is a solution of equation (3.4.1) in $L^p(G)$. If $\phi^* = \theta$, i.e., $K^{1/2}\psi^* = \theta$, then $\psi^* \in H_0$. But $\psi^* \in H_1$, hence $\psi^* = \theta$, a contradiction. Thus, we have $\psi^* \neq \theta$ and our theorem is proved. □

<u>Theorem 3.4.2</u>. If conditions (P_2), (Q_1), (Q_2), (Q_3'), (Q_4), and (Q_5) are satisfied and $k(x,y)$ has infinitely many eigenvalues, then the integral equation (3.4.1) has infinitely many nontrivial solutions in $L^p(G)$.

Proof. We will keep the notation used in the proof of Theorem 3.4.1. Since $k(x,y)$ has infinitely many eigenvalues, it follows from conclusion (ii) of Lemma 3.4.1 that H_1 is infinite-dimensional. We verify that the functional Ψ on H_1 satisfies all conditions (I_1), (I_3), (I_4) and (I_5) of Theorem 2.8 in Ambrosetti and Rabinowitz [1]. Conditions (I_1) and (I_3) have already been verified in the proof of Theorem 3.4.1. By (Q_5) we find

$$F(x,-u) = F(x,u), \quad x \in G, \quad -\infty < u < +\infty,$$

and hence, $\Psi(-\psi) = \Psi(\psi)$ for all $\psi \in H_1$, and (I_4) is satisfied. Finally, we verify (I_5). Suppose that (I_5) is not satisfied. Then, there exist a finite-dimensional subspace X of H_1 and $h_n^* \in X$ such that $\|h_n^*\| \to +\infty$ and

$$\Psi(h_n^*) \geq 0 \quad (n = 1,2,3,\ldots). \tag{3.4.32}$$

Letting $t_n = \|h_n^*\|$ and $\psi_n^* = t_n^{-1}h_n^*$, we have $h_n^* = t_n\psi_n^*$, $\|\psi_n^*\| = 1$ and $t_n \to +\infty$. Since X is finite-dimensional, $\{\psi_n^*\}$ contains a convergent subsequence. Without loss of generality, we may assume that ψ_n^* itself converges to some $\psi^* \in X$, i.e., $\|\psi_n^* - \psi^*\| \to 0$. Obviously, $\|\psi^*\| = 1$. The continuity of operator H implies

$$\|H\psi_n^* - H\psi^*\|_p = \left(\int_G |H\psi_n^*(x) - H\psi^*(x)|^p dx\right)^{1/p} \to 0 , \tag{3.4.33}$$

and as a result, $\{H\psi_n^*(x)\}$ contains a subsequence which converges to $H\psi^*(x)$ almost everywhere on G. Without loss of generality, we may also assume that $H\psi_n^*(x)$ itself converges to $H\psi^*(x)$ almost everywhere on G. Obviously, $H\psi^* \neq \theta$, since otherwise, $K^{1/2}\psi^* = H\psi^* = \theta$ implies that $\psi^* \in H_0$, and from $\psi^* \in X \subset H_1$ we get $\psi^* = \theta$, which contradicts the original $\|\psi^*\| = 1$. Setting $v_n = H\psi_n^*$ $(n = 1,2,3,\ldots)$, $v_0 = H\psi^*$,

$G_0 = \{x \in G \mid v_0(x) \neq 0, \quad v_n(x) \to v_0(x)\}$, and $a_0 = (\int_G [v_0(x)]^2 dx)^{1/2}$, we see that mes $G_0 > 0$ and $a_0 > 0$. By (Q_4), there exists a $R > 0$ such that

$$\frac{f(x,u)}{u} \geq \frac{8}{a_0^2}, \quad \forall\, x \in G, \quad |u| \geq R. \tag{3.4.34}$$

Similar to (3.4.24), we can deduce the following inequality

$$\int_G F(x, Hh_n^*)dx = \int_G F(x, t_n v_n)dx$$

$$\geq \frac{4}{a_0^2} t_n^2 \int_{D_n} [v_n(x)]^2 dx - M^*, \tag{3.4.35}$$

where $D_n = \{x \in G \mid t_n \mid v_n(x)| \geq R\}$ $(n = 1,2,3,\ldots)$ and

$$M^* = \left(\frac{4R^2}{a_0^2} + 2aR + \frac{2bR^p}{p}\right) \text{mes } G = \text{const.}$$

Using the Hölder inequality and observing (3.4.33), we find

$$\left|\left(\int_{D_n} [v_n(x)]^2 dx\right)^{1/2} - \left(\int_{D_n} [v_0(x)]^2 dx\right)^{1/2}\right|$$

$$\leq \left(\int_{D_n} [v_n(x) - v_0(x)]^2 dx\right)^{1/2}$$

$$\leq \left(\int_G [H\psi_n^*(x) - H\psi^*(x)]^2 dx\right)^{1/2}$$

$$\leq (\text{mes } G)^{\frac{1}{2} - \frac{1}{p}} \cdot \left(\int_G |H\psi_n^*(x) - H\psi^*(x)|^p dx\right)^{1/p} \to 0,$$

and hence, there exists a $N_1 > 0$ such that

$$\left(\int_{D_n} [v_n(x)]^2 dx\right)^{1/2} > \left(\int_{D_n} [v_0(x)]^2 dx\right)^{1/2} - \frac{a_0}{4}, \quad n > N_1. \tag{3.4.36}$$

Letting $D_n^* = \bigcap_{k=n}^{\infty} D_k$ and $D^* = \bigcup_{n=1}^{\infty} D_n^*$, we have $D_1^* \subset D_2^* \subset D_3^* \subset \ldots \subset D^*$, $D_n \supset D_n^*$ and mes $D_n^* \to$ mes D^*. Hence, there exists a $N_2 > 0$ such that

$$\left(\int_{D_n} [v_0(x)]^2 dx\right)^{1/2} \geq \left(\int_{D_n^*} [v_0(x)]^2 dx\right)^{1/2}$$

$$> \left(\int_{D^*} [v_0(x)]^2 dx\right)^{1/2} - \frac{a_0}{4}, \quad n > N_2. \qquad (3.4.37)$$

It is easy to see that $G_0 \subset D^*$, and therefore

$$\left(\int_{D^*} [v_0(x)]^2 dx\right)^{1/2} \geq \left(\int_{G_0} [v_0(x)]^2 dx\right)^{1/2} = a_0 . \qquad (3.4.38)$$

It follows from (3.4.36), (3.4.37), and (3.4.38) that

$$\left(\int_{D_n} v_n(x) \ {}^2 dx\right)^{1/2} > \frac{a_0}{\ }, \quad \forall \ n > N = \max\{N_1, N_2\} . \qquad (3.4.39)$$

By (3.4.35) and (3.4.39), we get

$$\Psi(h_n^*) \leq \frac{1}{2}|h_n^*|^2 - (t_n^2 - M^*) = -\frac{1}{2} t_n^2 + M^*, \ \forall \ n > N,$$

and hence, $\Psi(h_n^*) \to -\infty$ as $n \to \infty$, which contradicts (3.4.32). Thus, (I_5) is satisfied.

Now, by Theorem 2.8 and Corollary 2.9 in Ambrosetti and Rabinowitz [1], we know that Ψ has infinitely many nontrivial critical points ψ_n^{**} $(n = 1,2,3,\ldots)$ in H_1. As in the proof of Theorem 3.4.1, $\phi_n^* = H\psi_n^{**} \in L^p(G)$ $(n = 1,2,3,\ldots)$ are nontrivial solutions of equation (3.4.1). Moreover, it is easy to see that $\phi_n^* \neq \phi_m^*$ for $n \neq m$. In fact, if $\phi_n^* = \phi_m^*$ for some n and m with $n \neq m$, then $K^{1/2}(\psi_n^{**} - \psi_m^{**}) = \theta$, and therefore $\psi_n^{**} - \psi_m^{**} \in H_0$; but $\psi_n^{**} - \psi_m^{**} \in H_1$ and hence $\psi_n^{**} = \psi_m^{**}$, a contradiction. Thus, our proof is complete. □

It is easy to see from the proofs of Theorem 3.4.1 and Theorem 3.4.2 that condition (3.4.2) in (P_2) can be replaced by the following weaker condition:

K is completely continuous from $L^q(G)$ into $L^p(G)$
for some $p > 2$, where $1/p + 1/q = 1$. (3.4.40)

Besides, it is easy to find some elementary functions $f(x,u)$, which satisfy all conditions of Theorem 3.4.2; for example,

$$f(x,u) = \sum_{k=1}^{n} a_k u^{2k-1},\ n \geq 2,\ a_n > 0,\ a_1 < \frac{1}{\lambda_1}, \tag{3.4.41}$$

and

$$f(x,u) = \frac{u^7}{1 + u^4} - u^{1/3}. \tag{3.4.42}$$

We may take $p = 2n$ for function (3.4.41) and $p = 4$ for (3.4.42). We only verify condition (Q_2). For function (3.4.41), we have

$$\int_0^u f(x,v)dv = \sum_{k=1}^{n} \frac{a_k}{2k} u^{2k} \leq \sum_{k=1}^{n} \frac{a_k}{2n-1} u^{2k} = \frac{1}{2n-1} uf(x,u)$$

for sufficiently large $|u|$ and hence we may choose $\tau = 1/(2n - 1)$ with $0 < \tau \leq 1/3$. For function (3.4.42) we have

$$\int_0^u f(x,v)dv = \frac{u^4}{4} - \frac{1}{4} ln(1 + u^4) - \frac{3}{4} u^{4/3}$$

$$\leq \frac{u^4}{3} - \frac{u^4}{3(1 + u^4)} - \frac{1}{3} u^{4/3} = \frac{1}{3} uf(x,u)$$

for sufficiently large $|u|$ and therefore we may take $\tau = 1/3$.

We must also point out that Theorem 2.8 and Corollary 2.9 in Ambrosetti and Rabinowitz [1] only hold when the Banach space E is infinite-dimensional. For example, it is easy to verify that the function

$$f(u) = u^2(e^{-u^2} - \tfrac{1}{2}),\ \forall\ u \in R^1 = (-\infty,+\infty) \tag{3.4.43}$$

satisfies all conditions (I_1), (I_3), (I_4), and (I_5) (also (I_2)) of Theorem 2.8 in Ambrosetti and Rabinowitz [1], but f has only two non-trivial critical points. Thus, the hypothesis "$k(x,y)$ has infinitely many eigenvalues" in Theorem 3.4.2 cannot be omitted.

(b) Case of Quasi-Positive-Definite Kernels.

The following conditions are used:

(P_3) $k(x,y)$ is quasi-positive-definite and satisfies (3.4.2) for some $p > 2$.

(Q_3^*) There exist $\varepsilon_0 > 0$ with $\varepsilon_0 < -\lambda^*$ and $\delta > 0$ such that

$$\frac{f(x,u)}{u} \leq \frac{1}{\lambda^* + \varepsilon_0}, \quad x \in G, \quad 0 < |u| < \delta, \tag{3.4.44}$$

where λ^* denotes the largest negative eigenvalue of $k(x,y)$.

(Q_4^*) There exist $\eta > 0$ and $R > 0$ such that

$$\frac{f(x,u)}{u} \geq \eta, \quad x \in G, \quad |u| \geq R. \tag{3.4.45}$$

<u>Theorem 3.4.3</u>. If conditions (P_3), (Q_1), (Q_2), (Q_3^*), and (Q_4^*) are satisfied, then the integral equation (3.4.1) has at least one nontrivial solution in $L^p(G)$.

<u>Proof</u>. Let the sequence of eigenvalues of $k(x,y)$ be $\{-\lambda_0, -\lambda_1, \ldots, -\lambda_m, \lambda_{m+1}, \lambda_{m+2}, \ldots$, where $0 < \lambda_0 \leq \lambda_1 \leq \cdots \leq \lambda_m$, $\lambda_{m+1} \geq \lambda_{m+2} \geq \cdots > 0$ and hence $\lambda^* = -\lambda_0$. Choosing a number μ such that $0 < \mu < \lambda_0$ and setting $K_1 = KR_1$, where $R_1 = (K + \mu I)^{-1}$ and I denotes the identical operator, we see from Lemma 1.1 in Guo [17] that K_1 is also a linear integral operator generated by some L^2 kernel $k_1(x,y)$ with sequence of eigenvalues $\lambda_0/(\lambda_0 - \mu), \lambda_1/(\lambda_1 - \mu), \ldots, \lambda_m/(\lambda_m - \mu), \lambda_{m+1}/(\lambda_{m+1} + \mu), \lambda_{m+2}/(\lambda_{m+2} + \mu), \ldots$. Since all these eigenvalues are positive, $k_1(x,y)$ is a positive definite kernel with largest eigenvalue $\lambda_0^* = \lambda_0/(\lambda_0 - \mu)$. Moreover, from the proof of Theorem 1.1 in Guo [17] we know that K_1 is also a completely continuous operator from $L^q(G)$ into $L^p(G)$ with $1/p + 1/q = 1$.

Now, we consider the Hammerstein integral equation

$$\phi(x) = \int_G k_1(x,y) f_1(y, \phi(y))\, dy \ , \tag{3.4.46}$$

where

$$f_1(x,u) = u + \mu f(x,u) \ . \tag{3.4.47}$$

and prove that $k_1(x,y)$ and $f_1(x,u)$ satisfy all conditions of Theorem 3.4.1. In fact, (Q_1) for $f(x,u)$ implies (Q_1) for $f_1(x,u)$, and (P_2) is obviously satisfied for $k\ (x,y)$ by replacing (3.4.2) by (3.4.40). Since $f(x,u)$ satisfies (Q_2), we have

$$\begin{aligned} F_1(x,u) &= \int_0^u f_1(x,v)dv = \frac{u^2}{2} + \mu F(x,u) \\ &\le \frac{u^2}{2} + \mu\tau u f(x,u), \quad x \in G, \quad |u| \ge M \ . \end{aligned} \tag{3.4.48}$$

Choosing τ_1 such that $\tau < \tau_1 < 1/2$ and $(1/2 - \tau_1)/\mu(\tau_1 - \tau) < \eta$, we find by (Q_4^*)

$$\frac{uf(x,u)}{u^2} = \frac{f(x,u)}{u} \ge \eta > \frac{\frac{1}{2} - \tau_1}{\mu(\tau_1 - \tau)} \ , \ \forall \ x \in G, \quad |u| \ge R. \tag{3.4.49}$$

It follows from (3.4.48) and (3.4.49) that

$$\begin{aligned} F_1(x,u) &\le \frac{u^2}{2} + \mu\tau u f(x,u) < \tau_1 u^2 + \mu\tau_1 u f(x,u) \\ &= \tau_1 u f_1(x,u), \quad x \in G, \quad |u| \ge M_1 = \max\{M,R\} \ , \end{aligned}$$

and hence, (Q_2) is satisfied for $f_1(x,u)$. Since

$$1 - \frac{\mu}{\lambda_0 - \varepsilon_0} < 1 - \frac{\mu}{\lambda_0} = \frac{\lambda_0 - \mu}{\lambda_0} = \frac{1}{\lambda_0^*} < 1 \ ,$$

we can choose a sufficiently small positive number ε_1 such that $\varepsilon_1 < \lambda_0^*$ and

$$1 - \frac{\mu}{\lambda_0 - \varepsilon_0} < \frac{1}{\lambda_0^* + \varepsilon_1} < \frac{1}{\lambda_0^* - \varepsilon_1} < 1.$$

Thus, (Q_3^*) and (Q_4^*) imply

$$\frac{f_1(x,u)}{u} = 1 + \mu \frac{f(x,u)}{u} \leq 1 + \frac{\mu}{\lambda^* + \varepsilon_0}$$

$$= 1 - \frac{\mu}{\lambda_0 - \varepsilon_0} < \frac{1}{\lambda_0^* + \varepsilon_1}, \quad \forall \; x \in G, \quad 0 < |u| < \delta$$

and

$$\frac{f_1(x,u)}{u} = 1 + \mu \frac{f(x,u)}{u} \geq 1 + \mu\eta > 1 > \frac{1}{\lambda_0^* - \varepsilon_1}, \quad \forall \; x \in G, \quad |u| \geq R,$$

and therefore (Q_3') and (Q_4') are satisfied for $f_1(x,u)$.

Now, by Theorem 3.4.1 we know that the integral equation (3.4.46) has at least one nontrivial solution $\phi^*(x) \in L^p(G)$ and from the proof of Theorem 1.1 in Guo [17] we see that $\phi^*(x)$ is also a solution of equation (3.4.1). Our proof is complete.

Theorem 3.4.4. If conditions (P_3), (Q_1), (Q_2), (Q_3^*), (Q_4), and (Q_5) are satisfied and $k(x,y)$ has infinitely many eigenvalues, then the integral equation (3.4.1) has infinitely many nontrivial solutions in $L^p(G)$.

Proof. As before, we will keep the notation used in the proof of Theorem 3.4.3. We shall prove that $k_1(x,y)$ and $f_1(x,u)$ satisfy all conditions of Theorem 3.4.2. In fact, since the sequence $\{-\lambda_0, -\lambda_1, \ldots, -\lambda_m, \lambda_{m+1}, \lambda_{m\ 2} \ldots\}$ is infinite, the sequence

$$\frac{\lambda_0}{\lambda_0 - \mu}, \frac{\lambda_1}{\lambda_1 - \mu}, \ldots, \frac{\lambda_m}{\lambda_m - \mu}, \frac{\lambda_{m+1}}{\lambda_{m+1} + \mu}, \frac{\lambda_{m+2}}{\lambda_{m+2} + \mu}, \ldots$$

is also infinite, i.e., $k_1(x,y)$ has infinitely many eigenvalues. Moreover, (Q_4) and (Q_5) for $f(x,u)$ imply (Q_4) and (Q_5) for $f_1(x,u)$, respcetively:

$$\frac{f_1(x,u)}{u} = 1 + \mu \frac{f(x,u)}{u} \to +\infty \quad \text{as} \quad u \to \infty \quad \text{uniformly in} \quad x \in G,$$

$$f_1(x,-u) = -u + \mu f(x,-u) = -u - \mu f(x,u) = -f_1(x,u), \quad \forall \quad x \in G,$$

$$-\infty < u < +\infty \; .$$

Conditions (P_2), (Q_1), (Q_2), and (Q_3') for $k_1(x,y)$ and $f_1(x,u)$ have been verified already in the proof of Theorem 3.4.3. Hence, by virtue of Theorem 3.4.2, equation (3.4.46) has infinitely many nontrivial solutions ϕ_n^* $(n = 1,2,3,...)$ in $L^p(G)$ and from the proof of Theorem 1.1 in Guo [17] these ϕ_n^* $(n = 1,2,3,...)$ are also solutions of equation (3.4.1). Our theorem is proved. □

It is easy to point out some elementary functions $f(x,u)$, which satisfy all conditions of Theorem 3.4.4. For example, function (3.4.41) with condition $a_1 < 1/\lambda^*$ instead of $a_1 < 1/\lambda_1$, and function (3.4.42) are such elementary functions.

(c) *Applications.*

In this paragraph, we give some applications of Theorems 3.4.1 and 3.4.2 to the two-point boundary value problem:

$$\frac{-d^2u}{dx^2} = f(x,u) \ , \quad 0 \le x \le 1;$$
$$u(0) = u(1) = 0 \ . \tag{3.4.50}$$

<u>Theorem 3.4.5</u>. If $f(x,u)$ is continuous in $0 \le x \le 1$, $-\infty < u < +\infty$ and satisfies (Q_1) for some $p > 2$, (Q_2), (Q_3'), and (Q_4') with $G = [0,1]$ and $\lambda_1 = 1/\pi^2$, then problem (3.4.50) has at least one nontrivial solution in $C^2[0,1]$.

<u>Theorem 3.4.6</u>. If $f(x,u)$ is continuous in $0 \le x \le 1$, $-\infty < u < +\infty$ and satisfies (Q_1) for some $p > 2$, (Q_2), (Q_3'), (Q_4), and (Q_5) with $G = [0,1]$ and $\lambda_1 = 1/\pi^2$, then problem (3.4.50) has infinitely many nontrivial solutions in $C^2[0,1]$.

<u>Proofs of Theorems 3.4.5 and 3.4.6</u>. It is well known that the solution in $C^2[0,1]$ of problem (3.4.50) is equivalent to the solution in $C[0,1]$ of the Hammerstein integral equation

$$u(x) = \int_0^1 G(x,y)f(y,u(y))dy \ , \tag{3.4.51}$$

where $G(x,y)$ is the corresponding Green function:

$$G(x,y) = \begin{cases} x(1 - y), & x \leq y; \\ y(1 - x), & x > y. \end{cases}$$

It is also well known that the eigenvalues of $G(x,y)$ are $\{1/n^2\pi^2\}$ $(n = 1,2,3,\ldots)$ and the corresponding orthonormal eigenfunctions are $\{\sqrt{2} \sin n\pi x\}$ $(n = 1,2,3,\ldots)$. Hence, $G(x,y)$ is a positive-definite kernel with infinitely many eigenvalues. By virtue of the continuity of $G(x,y)$, we have

$$\int_0^1 \int_0^1 |G(x,y)|^p dxdy < +\infty \ ,$$

and hence $G(x,y)$ satisfies condition (P_2). It follows therefore from Theorem 3.4.1 and Theorem 3.4.2 that equation (3.4.51) has at least one nontrivial solution in $L^p[0,1]$ in the case of Theorem 3.4.5 and has infinitely many nontrivial solutions in $L^p[0,1]$ in the case of Theorem 3.4.6. It remains to show that every solution $u^*(x) \in L^p[0,1]$ of equation (3.4.51) must belong to $C[0,1]$. In fact, (Q_1) implies $f(x,u^*(x)) \in L^q[0,1]$, where $1/p + 1/q = 1$. It follows therefore by the continuity of $G(x,y)$ that the function

$$u^*(x) = \int_0^1 G(x,y)f(y,u^*(y))dy$$

belongs to $C[0,1]$. Our proofs are complete. □

Applying Theorem 3.4.6 to functions (3.4.41) and (3.4.42) in particular, we obtain the following two conclusions:

(A*) the boundary value problem

$$\begin{cases} -\dfrac{d^2u}{dx^2} = \sum\limits_{k=1}^{n} a_k u^{2k-1}, \quad 0 \le x \le 1; \\ u(0) = u(1) = 0 , \end{cases} \tag{3.4.52}$$

where $n \geq 2$, $a_n > 0$, and $a_1 < \pi^2$, has infinitely many nontrivial solutions in $C^2[0,1]$.

(B*) The boundary value problem

$$\begin{cases} -\dfrac{d^2u}{dx^2} = \dfrac{u^7}{1+u^4} - u^{1/3}, \quad 0 \le x \le 1 ; \\ u(0) = u(1) = 0 \end{cases} \tag{3.4.53}$$

has infinitely many nontrivial solutions in $C^2[0,1]$.

Since the sequence of orthonormal eigenfunctions $\{\sqrt{2}\sin n\pi x\}$ $(n = 1,2,3,\ldots)$ of $G(x,y)$ is not complete in $L^2[0,1]$, then $G(x,y)$ is not strictly positive-definite, and therefore $G(x,y)$ does not satisfy condition (P_1) by (3.4.10); moreover, condition (Q_3') is weaker than (Q_3) and function (3.4.41) satisfies (Q_3') but not (Q_3) when $a_1 \neq 0$. Hence, Theorems 3.4.5 and 3.4.6, and in particular, conclusions (A*) and (B*), cannot be deduced from the results (A) and (B) of Ambrosetti and Rabinowitz mentioned in the beginning of this section.

3.5 *Notes and Comments.*

Theorem 3. 1.1 to 3.1.11 are adapted from Guo [5], [18], [9] and Theorem 3.2.1 to 3.2.10 can be found in Guo [19], [11]. Notice that Theorem 3.2.7 is connected with Theorem 2 in Leggett and Williams [3], but both the method and conclusion are quite different. Theorem 3.3.1 to 3.3.3 are taken from Pazy and Rabinowitz [1], [2], while Theorem 3.3.4 is due to Guo [13]. Theorem 3.3.5 is obtained by Guo [14] and Guo and Qingyong [1] and is an improvement of a result found by Stuart [1]. Theorem 3.3.6 is due to Leggett and Williams [2], while Theorem 3.3.7 and 3.3.8 are taken from Guo's unpublished work. Finally, Theorem 3.4.1 to 3.4.6 are contained in Guo [16], and among these, Theorem 3.4.1 and Theorem 3.4.2 are improvements of Theorem 5.8 and Theorem 5.9 in Ambrosetti and Rabinowitz [1], respectively. For related results see Hively [1], Brezis and Browder [1], Thieme [1], Cahlon and Eskin [1], Miller [1], and Swick [1].

CHAPTER 4. APPLICATIONS TO NONLINEAR DIFFERENTIAL EQUATIONS

4.0 *Introduction.*

This chapter is devoted to the development of a theory of nonlinear differential equations in cones. We begin in Section 4.1 with the basic theory of differential inequalities, go on to prove the existence of maximal solutions, and then discuss comparison theorems. Section 4.2 treats flow invariance results and also considers the equivalence of quasimonotonicity property and the tangential condition that is necessary for the existence of solutions in closed sets. Sections 4.3 to 4.5 investigate various results utilizing the method of upper and lower solutions and the monotone iterative technique. A discussion of quasi-solutions and mixed monotone systems, including several special cases, is also considered here. Finally Section 4.6 is devoted to the method of cone-valued Lyapunov functions and its application to the theory of stability.

4.1 *Differential Inequalities.*

In order to develop the theory of differential inequalities relative to a cone K in a Banach space E, we need to introduce the concept of quasimonotonicity. A function f from E into E is said to be *quasimonotone nondecreasing* if $x \leq y$ and $\phi x = \phi y$ for some $\phi \in K_0^*$, then $\phi(f(x)) \leq \phi(f(y))$. If $E = R^n$ and $K = R_+^n$, the standard cone, the inequalities induced by cone K are componentwise and quasimonotonicity of f is reduced to

$$x \leq y \quad \text{and} \quad x_i = y_i, \quad 1 \leq i \leq n \quad \text{implies} \quad f_i(x) \leq f_i(y).$$

Theorem 4.1.1. Let K be a cone with nonempty interior K°. Assume that (i) $u, v \in C^1[R_+, E]$, $f \in C[R_+ \times E, E]$ and $f(t,x)$ is quasimonotone nondecreasing for each $t \in R_+$; (ii) $u'(t) - f(t,u(t)) < v'(t) - f(t,v(t))$,

$t \in (t_0, \infty)$. Then $u(t_0) < v(t_0)$ implies that $u(t) < v(t)$, $t \geq t_0$.

Proof. Suppose that the assertion of the theorem is false. Then there exists a $t_1 > t_0$ such that

$$v(t_1) - u(t_1) \in \partial K \quad \text{and} \quad v(t) - u(t) \in K^0, \quad t \in [t_0, t_1).$$

Thus, by Theorem 1.4.1 there exists a $\phi \in K_0^*$ with $\phi(v(t_1) - u(t_1)) = 0$. Setting

$$m(t) = \phi(v(t) - u(t)) ,$$

we see that $m(t) > 0$ for $t_0 \leq t < t_1$ and $m(t_1) = 0$. Consequently, $m'(t_1) \leq 0$. At $t = t_1$, we have

$$u(t_1) \leq v(t_1) \quad \text{and} \quad \phi(u(t_1)) = \phi(v(t_1)) .$$

Hence, using the quasimonotonicity of f and (ii), it follows that

$$m'(t_1) = \phi(v'(t_1) - u'(t_1)) > \phi(f(t_1, v(t_1)) - f(t_1, u(t_1))) \geq 0 .$$

This contradiction proves the theorem.

Utilizing this result, we can prove the existence of the maximal solution of

$$x' = f(t,x), \quad x(t_0) = x_0 , \tag{4.1.1}$$

relative to the cone K.

Theorem 4.1.2. Let K be a cone with nonempty interior K^0. Suppose that

(i) $f \in C[R_0, E]$ and $f(t,x)$ is quasimonotone nondecreasing in x for each $t \in [t_0, t_0 + a]$, where $R_0 = \{(t,x) : t \in [t_0, t_0 + a], \|x - x_0\| \leq b\}$;

(ii) f is uniformly continuous on R_0 (and hence we may assume that a,b are such that $\|f(t,x)\| \leq M$ on R_0);

(iii) $g \in C[[t_0,t_0+a] \times R_+,R]$ with $g(t,0) \equiv 0$ and the only solution of the scalar differential equation

$$u' = g(t,u), \quad u(t_0) = 0 \tag{4.1.2}$$

is the trivial solution;

(iv) $\alpha(A) - \alpha(\{x - hf(t,x) : x \in A\}) \leq hg(t,\alpha(A))$, for $h > 0$, $t \in [t_0,t_0+a]$, where A is a bounded subset of $B[x_0,b]$ and α is the measure of noncompactness.

Then there exists a maximal solution of (4.1.1) relative to K on $[t_0,t_0+\eta]$ where $\eta = \min(a,b/(M+1))$.

Proof. Let $y_0 \in K^0$ with $\|y_0\| = 1$. Consider the system

$$x' = f(t,x) + \frac{1}{n} y_0, \quad x(t_0) = x_0 + \frac{1}{n} y_0, \tag{4.1.3}$$

for each positive integer n. We have

$$\|f(t,x) + \frac{1}{n} y_0\| \leq \|f(t,x)\| + \frac{1}{n} \leq M + 1 \ .$$

Applying Theorem 2.7.2 in Lakshmikantham and Leela [1] with $V(t,A) = \alpha(A)$, we deduce that there exists a solution $x_n(t)$ of (4.1.3) for each n and a solution $x(t)$ of (4.1.1) on $[t_0,t_0+\eta]$. The equicontinuity of the family $\{x_n(t)\}$ follows easily. Noting that $\alpha(\{x_n(t_0)\}) = \alpha(\{x_0 + 1/n\, y_0\}) = 0$, because of the fact $x_0 + 1/n\, y_0$ converges to x_0, we can conclude, as in Theorem 2.7.2, that the set $\{x_n(t)\}$ is relatively compact for each $t \in [t_0,t_0+\eta]$. We can then apply Ascoli's theorem to obtain a subsequence of $\{x_n(t)\}$ which converges uniformly to a continuous function $r(t)$ on $[t_0,t_0+\eta]$. Using standard arguments it is easily seen that $r(t)$ is a solution of (4.1.1) on $[t_0,t_0+\eta]$. Let $x(t)$ be any solution (4.1.1) on $[t_0,t_0+\eta]$. Then

$$x'(t) - f(t,x(t)) = 0 < \frac{1}{n} y_0 = x_n'(t) - f(t,x_n(t))$$

and

$$x(t_0) = x_0 < x_0 + \frac{1}{n} y_0 = x_n(t_0) .$$

Applying Theorem 4.1.1, we get

$$x(t) \leq x_n(t), \quad t \in [t_0,t_0 + \eta]$$

and hence

$$x(t) \leq \lim_{n\to\infty} x_n(t) \equiv r(t), \quad t \in [t_0,t_0 + \eta] .$$

This shows that $r(t)$ is the desired maximal solution. The proof is complete. □

Having established the existence of the maximal solution of (4.1.1) relative to K, it is now easy to prove a comparison result. Clearly such a result generalizes well-known comparison theorems in finite dimensional spaces.

Theorem 4.1.3. Suppose that the assumptions of Theorem 4.1.2 are satisfied. Let $m \in C^1[[t_0,t_0 + \eta],E]$ and

$$m'(t) \leq f(t,m(t)), \quad t \in [t_0,t_0 + \eta].$$

If $m(t_0) \leq x_0$, then $m(t) \leq r(t)$, $t \in [t_0,t_0 + \eta]$, where $r(t)$ is the maximal solution of (4.1.1).

Proof. Let $x_n(t)$ be a solution of (4.1.3) as in Theorem 4.1.2. Note that $x(t_0) = x_0 < x_0 + 1/n\, y_0 = x_n(t_0)$ and

$$m'(t) - f(t,m(t)) \leq 0 < \frac{1}{n} y_0 = x_n'(t) - f(t,x_n(t)),$$

for $t \in [t_0,t_0 + \eta]$. By Theorem 4.1.1,

$$m(t) < x_n(t), \quad \text{for each } n \text{ and } t \in [t_0, t_0 + \eta].$$

Hence, $m(t) \le \lim_{n\to\infty} x_n(t) \equiv r(t)$, $t \in [t_0, t_0 + \eta]$.

Corollary 4.1.1. Let the hypotheses of Theorem 4.1.2 hold and let $f(t,0) \equiv 0$. Then the maximal solution $r(t)$ of (4.1.1) such that $r(t_0) = x_0 \in K$ remains in K for $t \in [t_0, t_0 + \eta]$.

The proof follows by choosing $m(t) \equiv 0$ in Theorem 4.1.3.

If the compactness type condition in Theorem 4.1.2 were strengthened, one could prove Theorems 4.1.2 and 4.1.3 without assuming the uniform continuity of f. One has to utilize the arguments of Theorem 2.5.3 in Lakshmikantham and Leela [1]. We give this below as an exercise.

Problem 4.1.1. Suppose that (i) $f \in C[R_0, E]$, $\|f(t,x)\| \le M$ on R_0 and $\eta = \min(a, b/(M + 1))$; (ii) $\alpha(f(I\times A)) \le g(\alpha(A))$ where $I = [t_0, t_0 + \alpha]$ and A is a bounded subset of $B[x_0, b]$; (iii) $g \in C[[0,2b], R]$,· $g(0) = 0$ and $u(t) \equiv 0$ is the only solution of (4.1.2); (iv) $f(t,x)$ is quasi-monotone nondecreasing in x for each $t \in [t_0, t_0 + \alpha]$. Then there exists a maximal solution on $[t_0, t_0 + \eta]$ for the problem (4.1.1) relative to the cone K, provided the interior of K is nonempty. If, further, $m \in C^1[[t_0, t_0 + \eta], E]$ and $m'(t) \le f(t, m(t))$, $t \in [t_0, t_0 + \eta]$, with $m(t_0) \le x_0$, then $m(t) \le r(t)$ on $[t_0, t_0 + \eta]$, where $r(t)$ is the maximal solution of (4.1.1) on $[t_0, t_0 + \eta]$. □

One could prove Theorems 4.1.2 and 4.1.3 under a different set of conditions without directly imposing compactness type conditions. We proceed by proving a useful convergence result.

Theorem 4.1.4. Assume that

(i) $f \in C[[t_0, t_0 + \alpha] \times E, E]$ and there exist constants M and $b > 0$ such that $\|f(t,x)\| \le M$ on R_0 and $Ma < b$, where $R_0 = \{(t,x) : t_0 \le t \le t_0 + a,\ \|x - x_0\| \le b\}$;

(ii) $f_n \in C[[t_0,t_0 + a] \times E,E]$ for $n \geq 1$ and the sequence $\{f_n\}$ converges uniformly to f on R_0;

(iii) for each $n \geq 1$, $x_n(t)$ is a solution of $x' = f_n(t,x)$, $x_n(t_0) = y_n$, existing on $[t_0,t_0 + a]$ such that $\lim_{n\to\infty} x_n(t) = x(t)$ whenever $\lim_{n\to\infty} y_n = x_0$.

Then the sequence $\{x_n(t)\}$ converges uniformly to $x(t)$ and $x(t)$ is a solution of (4.1.1).

Proof. By the uniform convergence of f_n to f, for $\varepsilon = 1$ there exists $N > 0$ such that $\|f_n(t,x) - f(t,x)\| \leq 1$ for all $(t,x) \in R_0$ and $n \geq N$. Thus, $\|f_n(t,x)\| \leq 1 + M$ for all $(t,x) \in R_0$ and $n \geq N$. For $\varepsilon > 0$, let $\delta = \varepsilon/(1 + M)$. If $t,s \in [t_0,t_0 + a]$ with $|t - s| < \delta$, then

$$\|x_n(t) - x_n(s)\| \leq \left| \int_s^t \|f_n(\tau,x_n(\tau))\| d\tau \right| \leq |t - s|(1 + M) < \varepsilon .$$

Therefore the family $\{x_n(t)\}_{n=N}^{\infty}$ is equicontinuous. Using this fact and the pointwise convergence of $\{x_n(t)\}$ to $x(t)$, it is easily seen that $x(t)$ is continuous on $[t_0,t_0 + a]$. Thus, $x(t)$ is uniformly continuous on $[t_0,t_0 + a]$, and it follows that $x_n(t)$ converges uniformly to $x(t)$ on $[t_0,t_0 + a]$. The standard arguments show that $x(t)$ is a solution of (4.1.1). □

Let us now consider, for each $n \geq 1$, the initial value problem (4.1.3), namely,

$$x_n' = f(t,x_n) + \frac{1}{n} y_0, \quad x_n(t_0) = x_0 + \frac{1}{n} y_0,$$

where $y_0 \in K^0$ with $\|y_0\| = 1$.

For convenience we list below some needed hypotheses:

(H_1) $f(t,x)$ is quasimonotone nondecreasing in x for each $t \in [t_0,t_0 + a]$;

(H_2) for each $n \geq 1$, a solution of (4.1.3) exists on $[t_0, t_0 + a]$;

(H_3) a solution of (4.1.1) exists on $[t_0, t_0 + a]$;

(H_4) K is a regular cone in E, that is, every monotonic bounded sequence has a limit;

(H_5) there exists a subsequence $\{x_{n_k}(t)\}$ of $\{x_n(t)\}$ which converges to a function $r(t)$.

The following theorem gives two sets of conditions guaranteeing the existence of the maximal solution of (4.1.1) and also a comparison result.

Theorem 4.1.5. Suppose that K is a cone with nonempty interior and that assumption (i) of Theorem 4.1.4 holds. Assume further that either

(a) (H_1), (H_2), (H_3), and (H_4) hold

or

(b) (H_1), (H_2), and (H_5) hold.

Then there exists a maximal solution of (4.1.1) on $[t_0, t_0 + a]$. Moreover, if either (a) or (b) holds, $y \in C[[t_0, t_0 + a], E]$ and

$$y'(t) \leq f(t, y(t)), \quad y(t_0) \leq x_0$$

then $y(t) \leq r(t, t_0, x_0)$ on $[t_0, t_0 + a]$, where $r(t, t_0, x_0)$ is the maximal solution of (4.1.1).

Proof. (I) Let

$$x_n'(t) = f(t, x_n(t)) + \frac{1}{n} y$$

and

$$x_{n+1}'(t) = f(t, x_{n+1}(t)) + \frac{1}{n+1} y_0.$$

Then

$$x_n'(t) > f(t, x_n(t)) + \frac{1}{n+1} y_0$$

and

$$x'_{n+1}(t) \leq f(t,x_{n+1}(t)) + \frac{1}{n+1} y_0$$

with $x_n(t_0) > x_{n+1}(t_0)$. By Theorem 4.1.1 and the fact that $f(t,x) + 1/n\ y_0$ is quasimonotone nondecreasing in x, $x_n(t) > x_{n+1}(t)$ for all $t \in [t_0, t_0 + a]$. Suppose that $x(t)$ is a solution of (4.1.1) on $[t_0, t_0 + a]$. Then

$$x'_n(t) = f(t,x_n(t)) + \frac{1}{n} y_0 > f(t,x_n(t))$$

and

$$x_n(t_0) = x + \frac{1}{n} y_0 > x_0 = x(t_0).$$

As before we are lead to $x_n(t) > x(t)$ on $[t_0, t_0 + a]$ for $n \geq 1$. Hence $x_n(t)$ is a decreasing sequence that is bounded below and, by the regularity of K, $x_n(t)$ converges. Clearly $f(t,x_n) + 1/n$ converges uniformly to $f(t,x)$ on $[t_0, t_0 + a]$. By Theorem 4.1.4, $x_n(t)$ converges uniformly to a solution $r(t)$ of (4.1.1). But

$$r(t) = \lim_{n\to\infty} x_n(t) \geq \lim_{n\to\infty} x(t) = x(t),$$

which implies that $r(t,t_0,x_0)$ is the maximal solution of (4.1.1) on $[t_0, t_0 + a]$.

(II) The subsequence $\{x_{n_k}(t)\}$ converges to a solution of (4.1.1) by Theorem 4.1.4. That this solution is maximal follows directly as in the proof of part (I).

(III) Consider a solution $x_n(t) = x_n(t,t_0,x_0 + 1/n\ y_0)$ of (4.1.3). Then $x'_n(t) > f(t,x_n(t))$ and $y'(t) \leq f(t,y(t))$ with $x_n(t_0) = x_0 + 1/n\ y_0 > x \geq y(t_0)$. By Theorem 4.1.1, $y(t) < x_n(t)$. Thus taking the limit either of the sequence used in (I) or the subsequence used in (II), we obtain the result. □

4.2 *Flow-Invariant Sets.*

Recall that the assumption that K^0 is nonempty (K^0 is the interior of a cone K) was crucial for the validity of the results of Section 4.1, particularly Theorem 4.1.1 on differential inequalities. Since there are cones whose interior is empty, such as cones consisting of nonnegative functions in L^p-spaces, it is interesting to prove Theorem 4.1.1 in cones whose interior is empty. Since the proof of such a result is closely related to the results on flow invariance, we begin by discussing the flow invariance of closed sets.

Let $F \subset E$ be a closed set and let $f \in C[R_+ \times E, E]$. Consider the differential equation

$$x' = f(t,x), \quad x(t_0) = x_0 \in F . \tag{4.2.1}$$

The set F is said to be *flow-invariant* with respect to f if every solution $x(t)$ of (4.2.1) on $[t_0,\infty)$ is such that $x(t) \in F$ for $t_0 \le t < \infty$. A set $A \subset E$ is called a *distance set* if to each $x \in E$ there corresponds a point $y \in A$ such that $d(x,A) = \|x - y\|$. A function $g \in C[R_+ \times R_+, R_+]$ is said to be a *uniqueness function* if the following holds: if $m \in C[R_+, R_+]$ is such that $m(t_0) \le 0$ and $D^+m(t) \le g(t,m(t))$, whenever $m(t) > 0$, then $m(t) \le 0$ for $t_0 \le t < \infty$.

We are now in a position to prove the following result on flow invariance of F.

Theorem 4.2.1. Let $F \subset E$ be closed and a distance set. Suppose further that

(i) $\lim\limits_{h \to 0} \frac{1}{h} d(x + hf(t,x), F) = 0, \quad t \in R_+, \quad x \in \partial F;$

(ii) $\|f(t,x) - f(t,y)\| \le g(t, \|x - y\|), \quad x \in E - F, \quad y \in \partial F,$ where g is a uniqueness function.

Then F is flow invariant with respect to f.

Proof. Let $x(t)$ be a solution of (4.2.1). Assume that $x(t) \in F$ for $t_0 \leq t < t_0 + a$, where $t_0 + a < \infty$ is maximal, that is, $x(t)$ leaves the set F at $t = t_0 + a$ for the first time. Let $x(t_1) \notin F$, $t_1 \in (t_0 + a, \infty)$ and let $y_0 \in \partial F$ be such that $d(x(t_1), F) = \|x(t_1) - y_0\|$. Set, for $t \in [t_0, \infty)$,

$$m(t) = d(x(t), F),$$

and

$$v(t) = \|x(t) - y_0\| .$$

For $h > 0$ sufficiently small, we have, letting $x = x(t_1)$,

$$\begin{aligned} m(t_1 + h) &\leq \|x(t_1 + h) - y_0 - hf(t_1, y_0)\| + d(y_0 + hf(t_1, y_0), F) \\ &\leq \|x + hf(t_1, x) - y_0 - hf(t_1, y_0)\| + \|x(t_1 + h) \\ &\qquad - x - hf(t_1, x)\| + d(y_0 + hf(t_1, y_0), F) \\ &\leq \|x - y_0\| + hg(t_1, \|x - y_0\|) + 0(h) . \end{aligned}$$

Since $m(t_1) = v(t_1) > 0$, we obtain

$$D^+ m(t_1) \leq g(t_1, m(t_1)) .$$

This implies, in view of the facts that g is a uniqueness function and $m(t_0) = 0$, that $m(t) \leq 0$, $t_0 \leq t < \infty$. This contradicts $d(x(t_1), F) = m(t_1) > 0$. The proof is complete. □

Remark 4.2.1. Observe that Theorem 4.2.1 is true when $F = K$, K a cone. Although K is not assumed to have nonempty interior, Theorem 4.2.1 requires that K must be a distance set. This is, however, a weaker assumption because the cones in L^p-spaces are distance sets whose interior is empty. Note also that every closed convex set in a reflexive Banach space is a distance set.

The requirement in Theorem 4.2.1 that F be a distance set can be dispensed with if the boundary condition (i) holds locally uniformly, that is, for every $(t,\bar{x}) \in (t_0,\infty) \times \partial F$ there exists a $\delta = \delta(t,\bar{x}) > 0$ such that

$$\lim_{h\to 0} \frac{1}{h} d(x + hf(t,x),G) = 0$$

uniformly for $\|x - \bar{x}\| < \delta$ and $x \in \partial F$. In this case, it is possible to choose a minimal sequence $\{x_n\}$ in F such that $\|x(t_1) - x_n\|$ approaches $d(x(t_1),F)$ as $n \to \infty$. This observation leads to the following.

Corollary 4.2.1. Let the assumptions of Theorem 4.2.1 hold except that F is not a distance set and the boundary condition (i) holds locally uniformly. Then F is flow invariant relative to f.

We shall next consider an interesting result which shows that equivalence of the quasimonotonicity property and the boundary condition.

If $f \in C[D,E]$, D being a subset of E, then

(a) the quasimonotonic nondecreasing property of f relative to the cone K means that whenever $y, z \in D$, $y \leq z$ and $\phi(z - y) = 0$ for some $\phi \in K_0^*$, we have $\phi(f(z) - f(y)) \geq 0$;

(b) the corresponding boundary condition takes the form

$$\lim_{h\to 0^+} \frac{1}{h} d(z - y + h(f(z) - f(y)),K) = 0, \quad y, z \in D, \quad z - y \in K.$$

Setting $x = z - y$, $g(x) = f(x + y) - f(y)$ and noting that $x \in D \cap K$, (a) and (b) can be expressed as

(P_1) $x \in K \cap D$, $\phi x = 0$ for some $\phi \in K_0^*$ implies $\phi(g(x)) \geq 0$;

(P_2) $\lim_{h\to 0^+} \frac{1}{h} d(x + hg(x),K) = 0$, $x \in K \cap D$, respectively. The following result shows that (P_1) and (P_2) are equivalent, which implies that the quasimonotonicity property and the boundary condition are equivalent.

Theorem 4.2.2. Let $g \in C[D,E]$, D being a subset of E. Then properties (P_1) and (P_2) are equivalent.

Proof. We shall first prove that (P_2) implies (P_1). Let $x \in K$ and $\phi x = 0$ for some $\phi \in K_0^*$. Because of (P_2), there exists, for each $h > 0$, a $z_h \in K$ such that

$$\|x + hg(x) - z_h\| = 0(h)$$

and hence

$$\phi(x + hg(x) - z_n) = 0(h).$$

Since

$$\phi(g(x)) = \frac{1}{h}\,\phi(x + hg(x)) = \frac{1}{h}\,\phi(x + hg(x) - z_n) + \frac{1}{h}\,\phi(z_h)$$

$$= 0(1) + \frac{1}{h}\,\phi(z_h)\ ,$$

and $\phi(z_h) \geq 0$, we obtain the desired inequality $\phi(g(x)) \geq 0$ by taking the limit as $h \to 0^+$.

To show that (P_1) implies (P_2); let us suppose that (P_2) does not hold. This means that

$$d(x + h_n g(x), K) \geq h_n \delta\ , \tag{4.2.2}$$

where $\delta > 0$ and $h_n \to 0$ as $n \to \infty$. Let A be the set

$$A = \{y \in E : \|y - (x + hg(x))\| < h\delta \text{ for some } h > 0\}.$$

We shall show that A is open, convex and that $A \cap K = \phi$. It is clear that A is open. To prove that A is convex; let $x_1, x_2 \in A$ and $y = \alpha x_1 + \beta x_2$ where $\alpha, \beta > 0$ with $\alpha + \beta = 1$. From the inequalities

$$\|x_i - x - h_i g(x)\| < h_i \delta, \quad i = 1,2,$$

we have

$$\|y - x - (\alpha h_1 + \beta h_2) g(x)\| < (\alpha h_1 + \beta h_2)\delta \ .$$

Hence $y \in A$ and the convexity of the set A is proved. To show that $A \cap K = \phi$, assume that $y \in A \cap K$, $\|y - x - hg(x)\| < h\delta$, where $h > 0$. We can also write $y = x + hz$ where $\|z - g(x)\| < \delta$. Fixing an h_n such that $0 < h_n < h$, consider the point $y_n = x + h_n z$. The point $y_n \in K$ because it is on the line connecting x and y and K is convex. It is also in A, since $\|y_n - x - h_n g(x)\| < h_n \delta$. But this contradicts (4.2.2).

We have thus far shown that the set A is open, convex and $A \cap K = \phi$. By Theorem 1.1.11 in Lakshmikantham and Leela [1], there is a hyperplane separating K and A, that is, there exists a $\phi \in K^*$ such that

$$\phi(K) \geq \alpha, \quad \phi(A) < \alpha.$$

Since x can be approximated by points from A, we have $\phi x = \alpha$, and since $\lambda x \in K$ for $\lambda \geq 0$, it follows that

$$\lambda\alpha = \phi(\lambda x) \geq \alpha \quad \text{for all} \quad \lambda \geq 0 \ .$$

This implies that $\alpha = 0$ and $\phi \in K_0^*$. The point $x + g(x)$ being in A, we get

$$\phi(g(x)) = \phi(x + g(x)) < 0,$$

which contradicts (P_1). The proof is therefore complete. □

We can now prove a differential inequality result without requiring that K^0 be nonempty.

<u>Theorem 4.2.3</u>. Let K be a cone in E. Assume that

(i) $u,v \in C^1[R_+,E]$, $f \in C[R_+ \times E,E]$ and $f(t,x)$ is quasimonotone nondecreasing in x relative to K, for each $t \in R_+$;

(ii) $u'(t) - f(t,u(t)) \leq v'(t) - f(t,v(t))$, $t \in (t_0,\infty)$;

(iii) $\|f(t,x) - f(t,y)\| \leq g(t,\|x - y\|)$, $x \in E - K$, $y \in \partial K$, where g is a uniqueness function;

(iv) K is a distance set.

Then $u(t_0) \leq v(t_0)$ implies that $u(t) \leq v(t)$, $t \geq t_0$.

<u>Proof</u>. The main idea is to reduce the theorem to Theorem 4.2.1. We note that $m(t) = v(t) = u(t)$ satisfies

$$m'(t) = H(t,m(t)), \quad t \in (t_0,\infty), \quad m(t_0) \geq 0,$$

where

$$H(t,w) = f(t,u(t) + w) - f(t,u(t)) + q(t),$$

$$q(t) = v'(t) - f(t,v(t)) - u'(t) + f(t,u(t)) \geq 0.$$

It is now enough to show that the cone K is flow invariant with respect to H, since flow invariance of K implies that $m(t) \in K$, $t \geq t_0$, which is equivalent to the conclusion of the theorem.

It is easy to see that $H(t,w)$ satisfies (iii). Let $w \in K$ and $\phi w = 0$ for some $\phi \in K_0^*$. Then

$$\phi(H(t,w)) = \phi(q(t)) + \phi(f(t,u(t) + w) - f(t,u(t))) \geq 0$$

since $\phi(q(t)) \geq 0$ and f is quasimonotone nondecreasing. This shows that H satisfies (P_1). By Theorem 4.2.2, (P_1) is equivalent to (P_2) and hence H satisfies (P_2). Now, applying Theorem 4.2.1, we get $m(t) \in K$, $t \geq t_0$, and the proof is complete. □

4.3 *Method of Upper and Lower Solutions.*

Consider the IVP

$$u' = f(t,u), \quad u(0) = u_0, \tag{4.3.1}$$

where $f \in C[J \times E, E]$, $J = [0,T]$ and E a real Banach space. Let $v,w \in C^1[J,E]$ be such that

$$v' \leq f(t,v), \quad w' \geq f(t,w) \quad \text{on } J. \tag{4.3.2}$$

Then v,w are said to be lower and upper solutions of (4.3.1) relative to a cone K in E defined in a natural way. In the special case when $E = R^n$ and $K = R^n_+$, one can prove the following result.

Theorem 4.3.1. Let $v,w \in C^1[J,R^n]$ be lower and upper solutions of (4.3.1) such that $v(t) \leq w(t)$ on J and let $f \in C[\Omega,R^n]$, where

$$\Omega = [(t,u) \in J \times R^n : v(t) \leq u \leq w(t), t \in J].$$

If f is quasimonotone nondecreasing in u, then there exists a solution $u(t)$ of (4.3.1) such that $v(t) \leq u(t) \leq w(t)$ on J, provided $v(0) \leq u(0) \leq w(0)$.

In fact, the conclusion of Theorem 4.3.1 is true without demanding f to be quasimonotone nondecreasing, which is restrictive. However, in this case, we need to strengthen the notion of upper and lower solutions of (4.3.1) as follows:

For each i,

$$\begin{aligned} v_i' &\leq f_i(t,\sigma) \quad \text{for all } \sigma \text{ such that } v(t) \leq \sigma \leq w(t) \\ &\qquad \text{and } v_i(t) = \sigma_i; \\ w_i' &\geq f_i(t,\sigma) \quad \text{for all } \sigma \text{ such that } v(t) \leq \sigma \leq w(t) \\ &\qquad \text{and } w_i(t) = \sigma_i. \end{aligned} \tag{4.3.3}$$

Theorem 4.3.2. Let $v,w \in C^1[J,R^n]$ with $v(t) \leq w(t)$ on J, satisfying (4.3.2). Let $f \in C[\Omega,R^n]$. Then there exists a solution u of (4.3.1) such that $v(t) \leq u(t) \leq w(t)$ on J provided $v(0) \leq u(0) \leq w(0)$.

Since the assumptions of Theorem 4.3.1 imply the assumptions of Theorem 4.3.2 it is enough to prove Theorem 4.3.2.

Proof. Consider $P : J \times R^n \to R^n$ defined by

$$P_i(t,u) = \max\{v_i(t), \min[u_i, w_i(t)]\}, \quad \text{for each } i.$$

Then $f(t,P(t,u))$ defines a continuous extension of f to $J \times R^n$, which is also bounded since f is bounded on Ω. Therefore $u' = f(t,P(t,u))$ has a solution u on J with $u(0) = u_0$. Let us show that $v(t) \leq u(t) \leq w(t)$ and therefore a solution of (4.3.1). For $\varepsilon > 0$ and $e = (1,\ldots,1)$, consider $w_\varepsilon(t) = w(t) + \varepsilon(1 + t)e$ and $v_\varepsilon(t) = v(t) - \varepsilon(1 + t)e$. We have $v_\varepsilon(0) < u_0 < w_\varepsilon(0)$. Suppose that $t_1 \in J$ is such that $v_\varepsilon(t) < u(t) < w_\varepsilon(t)$ in $[0,t_1)$ but $u_j(t_1) = w_{\varepsilon j}(t_1)$. Then we have $v(t_1) \leq p(t_1,u(t_1)) \leq w(t_1)$ and $P_j(t_1,u(t_1)) = w_j(t_1)$, hence

$$w_j'(t_1) \geq f_j(t_1,P(t_1,u(t_1))) = u_j'(t\ ),$$

which implies $u_j'(t_1) < w'_j(t_1)$, contradicting $u_j(t) < w_{\varepsilon j}(t)$ for $t < t_1$. Therefore $v_\varepsilon(t) < u(t) < w_\varepsilon(t)$ in J. Now $\varepsilon \to 0$ yields $v(t) \leq u(t) \leq w(t)$, and the proof is complete. □

Next we use the idea of differential inequalities to give a different proof of Theorem 4.3.2 which is interesting in itself. For this proof we need to satisfy a Lipschitz condition of the form

$$|f_i(t,x) - f_i(t,y)| \leq L_i \sum_{i=1}^{n} |x_i - y_i| \ .$$

Proof. We first assume that v,w in (4.3.3) satisfy strict inequalities and also $v(0) < u(0) < w(0)$ and prove the conclusion for strict inequalities. If the conclusion is false there exists a $t_1 > 0$ and i, $1 \le i \le n$ such that $v(t_1) \le u(t_1) \le w(t_1)$ and either $v_i(t_i) = u_i(t_1)$ or $u_i(t_1) = w_i(t_1)$. Then

$$f_i(t_1,u(t_1)) = u_i'(t_1) \le v_i'(t_1) < f_i(t_1,\sigma_1,\ldots,v_i(t_1),\ldots\sigma_n)$$
$$= f_i(t_1,u(t_1))$$

or

$$f_i(t_1,u(t_1)) = u_i'(t_1) \ge w_i'(t_1) > f_i(t_1,\sigma_1,\ldots w_i(t_1),\ldots\sigma_n)$$
$$= f_i(t_1,u(t_1)),$$

which leads to a contradiction. In order to prove the result for nonstrict inequalities we consider

$$\tilde{w}_i(t) = w_i(t) + \varepsilon e^{(n+1)L_i t}, \quad \tilde{v}_i(t) = v_i(t) - \varepsilon e^{(n+1)L_i t},$$

where $\varepsilon > 0$ is sufficiently small. Further let

$$P_i(t,u) = \max\{v_i(t),\min(u_i,w_i(t))\} \quad \text{for each } i.$$

Then it is clear that if $\tilde{\sigma}$ is such that $\tilde{v}(t) \le \tilde{\sigma} \le \tilde{w}(t)$ and $\tilde{\sigma}_i = \tilde{w}_i(t)$, it follows that $\sigma = P(t,\tilde{\sigma})$ satisfies $v(t) \le \sigma \le w(t)$ and $\sigma_i = w_i(t)$. Hence using the Lipschitz condition it follows that

$$\tilde{w}_i'(t) = w_i'(t) + \varepsilon(n + 1)L_i e^{(n+1)L_i t} \ge f_i(t,\sigma) + \varepsilon(n + 1)L_i e^{(n+1)L_i t}$$
$$\ge f_i(t,\tilde{\sigma}) + \varepsilon e^{(n+1)L_i t} > f_i(t,\tilde{\sigma})$$

for all $\tilde{\sigma}$ such that $\tilde{v}(t) \le \tilde{\sigma} \le \tilde{w}(t)$ and $\tilde{\sigma}_i = \tilde{w}_i(t)$. Here we have used the fact that $|\tilde{\sigma}_j - P_j(t,\sigma)| \le \varepsilon e^{(n+1)L_i t}$ for each j. Similarly we get $\tilde{v}_i < f_i(t,\tilde{\sigma})$ for all $\tilde{\sigma}$ such that $\tilde{v}(t) \le \tilde{\sigma} \le \tilde{w}(t)$ and $\tilde{\sigma}_i = \tilde{v}_i(t)$.

Since $\tilde{v}(0) < u_0 < \tilde{w}(0)$ we can conclude from the previous argument that $\tilde{v}(t) < u(t) < \tilde{w}(t)$, $t \geq 0$. As ε is arbitrary, the result follows letting $\varepsilon \to 0$. □

We observe that the proofs of Theorems 4.3.1 and 4.3.2 depend crucially on the modification of f, that is $f(t,P(t,u))$ where $P(t,u) = \max[v(t),\min(u,w(t))]$. Clearly this modification makes sense only when $K = R^n_+$.

If K is an arbitrary cone, the inequalities (4.3.3) need not be changed. On the other hand, the inequalities can be formulated in terms of linear functionals from K^*, namely for $\phi \in K^*$,

$$\phi(v' - f(t,\sigma)) \leq 0 \quad \text{for all } \sigma \text{ such that } v(t) \leq \sigma \leq w(t)$$
$$\text{and } \phi(v(t) - \sigma) = 0,$$

$$\phi(w' - f(t,\sigma)) \geq 0 \quad \text{for all } \sigma \text{ such that } v(t) \leq \sigma \leq w(t)$$
$$\text{and } \phi(w(t) - \sigma) = 0. \tag{4.3.4}$$

This version of inequalities (4.3.3) allows us to consider cones K other than the standard cone. The question is whether Theorem 4.3.1 and 4.3.2 hold even when K is an arbitrary cone.

Consider the example in R^3. Let $K = [u \in R^n : (u_1^2 + u_2^2)^{1/2} \leq u_3]$. Take $v = (0,0,0)$, $w = (2,0,2)$, $f_1 = f_3 = 0$ and

$$f_2 = \begin{cases} u_1 & \text{if } u_1 \in [0,1], \\ 2 - u_1 & \text{if } u_1 \in [1,2], \\ 0 & \text{otherwise}. \end{cases}$$

The solution of (4.3.1) with $u_0 = (1,0,1)$ is $u(t) = (1,t,1)$ which does not remain in the sector $[v,w]$. Note also that f is Lipschitzian. To verify the inequalities (4.3.3), we see that they now have the form

$$0 \le \phi(f(z)) \quad \text{for} \quad 0 \le z \le w, \quad \phi \in K^* \quad \text{and} \quad \phi(z) = 0,$$

$$0 \ge \phi(f(z)) \quad \text{for} \quad 0 \le z \le w, \quad \phi \in K^* \quad \text{and} \quad \phi(z) = \phi(w) \ . \tag{4.3.5}$$

Since $K \cap (w - K) = [\lambda w : 0 \le \lambda \le 1]$, all z's with $0 \le z \le w$ have the form $z = \lambda w$. For $z = 0$ and $z = w$, (4.3.5) are true if we take into account that $f(0) = f(w) = 0$. Let us choose now $z = \lambda w$, $0 < \lambda < 1$. For such a z, we have either $\phi(z) = 0$ or $\phi(z) = \phi(w)$. In both cases this implies $\phi(w) = 0$ and (4.3.5) follows from

$$\phi(f(\lambda w)) = 0 \quad \text{for} \quad 0 < \lambda < 1, \quad \phi \quad K^*, \quad \phi(w) = 0, \tag{4.3.6}$$

which has been proved already.

Let $\phi = (\phi_1,\phi_2,\phi_3) \in K^*$ and $\phi(w) = 0$. Since $w = (2,0,2)$, we see that

$$\phi_1 + \phi_2 = 0. \tag{4.3.7}$$

For $u = (0,0,1) \in K$, $\phi_3 = \phi(u) \ge 0$ and for $\bar{u} = (-\phi_1,-\phi_2,\sqrt{\phi_1^2 + \phi_2^2}) \in K$, we have

$$-\phi_1^2 - \phi_2^2 + \phi_3\sqrt{\phi_1^2 + \phi_2^2} = \phi(\bar{u}) \ge 0.$$

With $\phi_3 \ge 0$, we obtain $\phi_3 \ge \sqrt{\phi_1^2 + \phi_2^2}$ and from (4.3.7) we get $\phi_2 = 0$. Hence (4.3.6) holds since $f_1(\lambda w) = f_3(\lambda w) = 0$. □

Thus we see Theorem 4.3.2 may not be valid even in R^3 if K is arbitrary. In the next section we shall prove Theorems 4.3.1 and 4.3.2 for an arbitrary cone using monotone iterative techniques.

4.4 *Monotone Iterative Technique.*

For any $v_0,w_0 \in C[J,E]$ such that $v_0(t) \le w_0(t)$ on J where $J = [0,T]$, we define the conical segment $[v_0,w_0] = [u \in C[J,E] : v_0 \le u \le w_0]$.

Let us consider the IVP

$$u' = f(t,u), \quad u(0) = u_0, \tag{4.4.1}$$

where $f \in C[J \times E, E]$. Let us list the following assumptions for convenience.

(A1) For any bounded set B in E

$\alpha(f(I \times B)) \leq L\alpha(B)$, for some $L > 0$,

where α denotes the Kuratowski's measure of noncompactness;

(A2) $v_0, w_0 \in C^1[J,E]$ with $v_0(t) \leq w_0(t)$ on J such that

$v_0' \leq f(t,v_0)$,

$w_0' \geq f(t,w_0)$ on J;

(A3) $f(t,u) - f(t,v) \geq -M(u - v)$ whenever $u \geq v$ and $u,v \in [v_0,w_0]$ for some $M > 0$.

We note that when (A3) holds, f is quasimonotone.

In order to develop the monotone iterative technique, we need to consider the linear IVP

$$u' + Mu = \sigma(t), \quad u(0) = u_0, \tag{4.4.2}$$

where $\sigma(t) = f(t,\eta(t)) + M\eta(t)$ and $\eta \in C[I,E]$ such that $v_0(t) \leq \eta(t) \leq w_0(t)$ on J. Clearly the linear IVP (4.4.2) possesses a unique solution on J.

For each $\eta \in C[I,E]$ such that $v_0(t) \leq \eta(t) \leq w_0(t)$ on J, we define the mapping A by $A\eta = u$, where u is the unique solution of (4.4.2) corresponding to η. The following result concerning mapping A holds.

Lemma 4.4.1. Suppose that assumptions (A1), (A2), and (A3) hold; then

(i) $v_0 \le Av_0$ and $w_0 \ge Aw_0$;

(ii) A is monotone on $[v_0, w_0]$, that if $\eta_1, \eta_2 \in [v_0, w_0]$ with $\eta_1 \le \eta_2$ then $A\eta_1 \le A\eta_2$.

Proof. (i) Suppose that $Av_0 = v_1$. Set $p(t) = \phi[v_1(t) - v_0(t)]$ so that $p(0) \ge 0$, where $\phi \in K^*$. Then

$$p' \ge \phi[f(t,v_0) - M(v_1 - v_0) - f(t,v_0)] = -Mp$$

in view of (A2). As a result we have $p(t) \ge p(0)e^{-Mt} \ge 0$ on J. Since $\phi \in K^*$ is arbitrary, this implies $v_1 \ge v_0$ on J proving $v_0 \le Av_0$. Similarly we can show that $w_0 \ge Aw_0$.

To prove (ii), let $\eta_1, \eta_2 \in C[J,E]$ such that $\eta_1 \le \eta_2$ on J and suppose that $A\eta_1 = u_1$, $A\eta_2 = u_2$. We set $p(t) = \phi[u_2(t) - u_1(t)]$ so that $p(0) \ge 0$, where, as before, $\phi \in K^*$. It then follows by using (A3) that

$$p' = \phi[f(t,\eta_2) - M(u_2 - \eta_2) - f(t,\eta_1) + M(u_1 - \eta_1)]$$

$$\ge \phi[-M(\eta_2 - \eta_1) - M(u_2 - \eta_2) + M(u_1 - \eta_1)] = -Mp.$$

Consequently $p(t) \ge p(0)e^{-Mt} \ge 0$ on J and this proves $A\eta_1 \le A\eta_2$. The proof of the lemma is complete. □

In view of Lemma 4.4.1, we can define the sequences $\{v_n\}, \{w_n\}$ as follows:

$$v_n = Av_{n-1} \quad \text{and} \quad w_n = Aw_{n-1}.$$

It is easy to see that $\{v_n\}, \{w_n\}$ are monotone sequences such that $v_n \le w_n$ and $v_n, w_n \in [v_0, w_0]$. We shall now show that there exist subsequences of $\{v_n\}, \{w_n\}$ which converge uniformly on I.

Lemma 4.4.2. Under the assumptions of Lemma 4.4.1, if the cone K is normal, then the sequences $\{v_n\},\{w_n\}$ are uniformly bounded, equicontinuous and relatively compact on I.

Proof. Since the cone K is assumed to be normal, it follows from v_n, $w_n \in [v_0, w_0]$ that $\{v_n\},\{w_n\}$ are uniformly bounded. This implies the equicontinuity of the sequences by using standard estimates and the fact that f maps bounded sets into bounded sets which is a consequence of (A1). Now we set $B(t) = \{v_n(t)\}_{n=}^{\infty}$ so that $B'(t) = \{v_n'(t)\}_{n=}^{\infty}$ and $m(t) = \alpha(B(t))$. Using the standard arguments as in Lakshmikantham and Leela [1], we get

$$D^- m(t) \le \alpha\left(\left\{\frac{v_n(t) - v_n(t-h)}{h}\right\}_{n=0}^{\infty}\right) \le \alpha(\overline{\text{conv}}(\{v_n'(t)\}_{n=0}^{\infty})).$$

Thus we have

$$D^- m(t) \le \lim_{h\to 0^+} \alpha\left(\bigcup_{J_h} B'(x)\right), \quad \text{where} \quad J_h = [t-h,t] \subset J.$$

Evidently

$$\begin{aligned}\alpha\left(\bigcup_{J_h} B'(S)\right) &\le \alpha\left(\bigcup_{J_h} \{f(t,v_{n-1}(t)\}_{n=1}^{\infty}\right) + 2M\alpha\left(\bigcup_{J_h} \{v_n(t)\}_{n=0}^{\infty}\right) \\ &\le \alpha\left(\left\{f\left(J, \bigcup_{J_h} B(s)\right)\right\}\right) + 2M\alpha\left(\bigcup_{J_h} B(s)\right) \\ &\le (L + 2M)\alpha\left(\bigcup_{J_h} B(s)\right).\end{aligned}$$

The equicontinuity of $v_n(t)$ now yields

$$D^- m(t) \le (L + 2M)m(t), \quad t \in J.$$

Since $m(0) = \alpha(\{v_n(0)\}_{n=0}^{\infty}) = \alpha(u_0, v_0(0)) = 0$, it is immediate that $m(t) \equiv 0$ on J, which implies the relative compactness of the sequence $\{v_n(t)\}$ for each $t \in I$. Similarly $\{w_n(t)\}$ is relatively compact for each $t \in J$. The proof of the lemma is complete. □

We now apply Ascoli's theorem to the sequences $\{v_n\},\{w_n\}$ to obtain subsequences $\{v_{n_k}\},\{w_{n_k}\}$ which converge uniformly on J. Since the sequences $\{v_n\},\{w_n\}$ are monotone, this then shows that the full sequences converge uniformly and monotonically to continuous functions, that is, $\lim_{n\to\infty} v_n(t) = \rho(t)$ and $\lim_{n\to\infty} w_n(t) = r(t)$ on I. It then follows easily from (4.4.2) that $\rho(t)$ and $r(t)$ are solutions of IVP (4.4.1) on J.

Finally we show that $\rho(t)$, $r(t)$ are minimal and maximal solutions of (4.4.1). To this end, let $u(t)$ be any solution of (4.4.1) on J such that $u \in [v_0, w_0]$. Assume that $v_n \leq u \leq w_n$ on J. Set $p(t) = \phi[u(t) - v_{n+1}(t)]$, so that $p(0) = 0$, where, as before, $\phi \in K^*$. Then by (A3) and the assumption $v_n \leq u$, we have

$$p' = \phi[f(t,u) - f(t,v_n) + M(v_{n+1} - v_n)]$$

$$\geq [-M(u - v_n) + M(v_{n+1} - v_n)] = -Mp.$$

This implies $v_{n+1} \leq u$ on J. Similarly, we can show $u \leq w_{n+1}$ on J. Since $u \in [v_0, w_0]$, we have, by induction, $v_n \leq u \leq w_n$ on J for all n. Thus we obtain, taking the limit as $n \to \infty$, $\rho(t) \leq u(t) \leq r(t)$ on J, proving $\rho(t)$, $r(t)$ are minimal and maximal solutions of (4.4.1) on J. We have therefore proved the following result which corresponds to Theorem 4.3.1. □

Theorem 4.4.1. Let the cone K be normal and assumptions (A1), (A2), and (A3) hold. Then there exist monotone sequences $\{v_n\},\{w_n\}$ which converge uniformly and monotonically to the minimal and maximal solutions $\rho(t)$, $r(t)$, respectively, of the IVP on $[v_0, w_0]$. That is, if u is any solution of (4.4.1) in $[v_0, w_0]$, then

$$v_0 \leq v_1 \leq \cdots \leq v_n \leq \rho \leq u \leq r \leq w_n \leq \cdots \leq w_1 \leq w_0 \quad \text{on } J.$$

Corollary 4.4.1. If the solutions of IVP (4.4.1) are unique, then the assumptions of Theorem 3.1 imply that $\rho(t) = u(t) = r(t)$ on J.

Remark 4.4.1. If f is quasimonotone relative to K where K is a solid cone, the existence of extremal solutions is given in Section 4.1. On the other hand, when K is not assumed to be solid and function f maps $I \times K$ into E, existence of extremal solutions is also known. See Deimling and Lakshmikantham [1].

Remark 4.4.2. Suppose that $E = \mathbb{R}^n$ and $K = \mathbb{R}^n_+$, the standard cone. Let $f(t,u)$ be quasimonotone nondecreasing in u for each $t \in I$, that is, $v \leq u$ and $u_i = v_i$ implies $f_i(t,v) \leq f_i(t,u)$. Suppose further that for each i, $i = 1,2,\ldots,n$,

$$f_i(t,v_1,\ldots,u_i,\ldots,v_n) - f_i(t,v_1,\ldots,v_i,\ldots,v_n) \geq -M(u_i - v_i),$$

where $v_{0i}(t) \leq v_i \leq u_i \leq w_{0i}(t)$. Then the conclusion of Theorem 4.4.1 is valid, provided (A2) holds. In this case, the forgoing two conditions on f imply

$$f(t,u) - f(t,v) \geq -M(u - v) \quad \text{where} \quad v_0 \leq v \leq u \leq w_0,$$

which corresponds to (A3). Thus when we assume the last condition, the quasimonotonicity of f is subsumed in it. Of course, one does not need (A1) when $E = \mathbb{R}^n$.

We shall next consider a result corresponding to Theorem 4.3.2. We begin listing the following assumptions.

(A4) $\|f(t,u_1) - f(t,u_2)\| \leq L\|u_1 - u_2\|$, $t \in J$, $u_1, u_2 \in [v_0,w_0]$;

(A5) $v_0,w_0 \in C^1[I,E]$ with $v_0(t) \leq w_0(t)$ on J such that there is an $M > 0$ satisfying

$$\phi[v_0' - f(t,\sigma) + M(v_0 - \sigma)] \leq 0, \quad \phi[w_0' - f(t,\sigma) + M(w_0 - \sigma)] \geq 0,$$

for all $\sigma \in [v_0,w_0]$ and $\phi \in K^*$.

Remark. If f is assumed to be uniformly continuous, then condition (A1) is superfluous. For, in that case,

$$(f(I \times B)) = \max_J \alpha(f(t,B)).$$

Remark. If, in addition, σ in (A5) is such that for every $\phi \in K^*$, $\phi(v_0(t) - \sigma) = 0$ and $\phi(w_0(t) - \sigma) = 0$, then condition (A5) reduces to condition (4.3.4). Suppose now that f satisfies (A3) then it is easy to show that condition (A2) implies (A5).

For any $\eta \in C[I,E]$ such that $v_0(t) \leq \eta(t) \leq w_0(t)$ on J, define the mapping A by $A\eta = u$, where $u = u(t)$ is the unique solution of (4.4.2) corresponding to η. Concerning the mapping A we have the following lemma.

Lemma 4.4.3. Suppose that assumptions (A1) and (A5) hold. Then A maps the sector $[v_0,w_0]$ into itself.

Proof. Let $\eta \in C[J,E]$ be such that $\eta \in [v_0,w_0]$ and let $u = A\eta$. For any $\phi \in K^*$, set $p(t) = \phi[u(t) - v_0(t)]$ so that $p(0) \geq 0$. Then for all $\sigma \in [v_0,w_0]$,

$$p' \geq \phi[f(t,\eta) - M(u - \eta) - f(t,\sigma) + M(v_0 - \sigma)],$$

in view of (A5). Choosing $\sigma = \eta$, we have $p' \geq -Mp$ which implies $p(t) \geq p(0)e^{-Mt} \geq 0$ on J. This proves $v_0(t) \leq u(t)$ on J. A similar argument shows that $u(t) \leq w_0(t)$ on J. Hence $u = A\eta$ $[v_0,w_0]$. Since η is arbitrary the proof is complete. □

In view of Lemma 4.4.3 we can define the sequence $u_n = Au_{n-1}$ with $u_0 = v_0$ or w_0 satisfying $u_n \in [v_0,w_0]$ on J.

By Lemma 4.4.2, we can conclude by Ascoli's theorem that there exists uniformly convergent subsequences of $\{u_n\}$. Suppose that $u_n(t) - u_{n-1}(t) \to 0$ as $n \to \infty$, then it is clear from the definition of $\{u_n\}$ that the limit of any subsequence is the unique solution of (4.4.1) on J. It then follows that a selection of a subsequence is unnecessary and the full sequence $\{u_n(t)\}$ converges uniformly to the unique solution $u(t)$ on J such that $u(t) \in [v_0, w_0]$ on J. Thus it is sufficient to prove that $m(t) = 0$ on J, where $m(t) = \limsup_{n\to\infty} \|u_n(t) - u_{n-1}(t)\|$. To this end, we have the following lemma.

Lemma 4.4.4. Let K be a normal cone and let assumptions (A1), (A4), and (A6) hold. Then $m(t) = 0$ on J.

Proof. Since f maps bounded sets into bounded sets, we let $\|f(t,u)\| \le N$ for $t \in J$ and $u \in [v_0, w_0]$. Then for $t_1, t_2 \in J$,

$$\|u_n(t_1) - u_{n-1}(t_1)\| \le \|u_n(t_2) - u_{n-1}(t_2)\| + 2N|t_1 - t_2|$$

$$\le m(t_2) + 2N|t_1 - t_2| + \varepsilon$$

for large n, given $\varepsilon > 0$. Hence $m(t_1) \le m(t_2) + 2N|t_1 - t_2| + \varepsilon$. Since t_1, t_2 can be interchanged and $\varepsilon > 0$ is arbitrary, we obtain

$$|m(t_1) - m(t_2)| \le 2N|t_1 - t_2|$$

which proves that $m(t)$ is continuous on J. Now (A4) yields

$$\|u_{n+1}(t) - u_n(t)\| \le \int_0^t [\|f(s,u_n(s)) - f(s,u_{n-1}(s))\|$$

$$+ M\|u_n(s) - u_{n-1}(s)\| + M\|u_{n+1}(s) - u_n(s)\|]ds$$

$$\le \int_0^t [(L + M)\|u_n(s) - u_{n-1}(s)\| + M\|u_{n+1}(s) - u_n(s)\|]ds .$$

For a fixed $t \in (0,T]$, there is a sequence of integers $n_1 < n_2 < \ldots$, such that $\|u_{n+1}(t) - u_n(t)\| \to m(t)$ as $n = n_k \to \infty$ and that $m^*(s) = \lim_{n=n_k\to\infty} \|u_n(s) - u_{n-1}(s)\|$ exists uniformly on J. It therefore follows because of the fact $m^*(s) \le m(s)$

$$m(t) \leq (L + 2M) \int_0^t m(s)ds \quad \text{on} \quad J,$$

which implies that $m(t) \leq m(0)e^{(L+2M)t}$, $t \in J$. Since $m(0) = 0$, we have $m(t) = 0$ on J proving the lemma.□

We have therefore proved the following result.

Theorem 4.4.2. Assume that the cone K is normal and that conditions (A1), (A4), and (A5) are satisfied. Then there exists a unique solution $u(t)$ of (4.4.1) on J such that $v_0(t) \leq u(t) \leq w_0(t)$ on J provided $v_0(0) \leq u_0 \leq w_0(0)$.

Corollary 4.4.2. Let $E = R^n$ and let (A4), (A5) hold. Then there exists a unique solution of (4.4.1) on J such that $u(t) \in [v_0, w_0]$, provided that $v_0(0) \leq u_0 \leq w_0(0)$.

4.5 *Method of Upper and Lower Quasi-solutions.*

We shall consider, in this section, a general situation where f in (4.4.1) admits a mixed quasimonotone property. To define appropriate classes of upper and lower solutions suitable to our needs, we shall suppose that f admits a decomposition of the form

$$f(t,u) = f_0(t,u) + f_1(t,u) + f_2(t,u) \tag{4.5.1}$$

where $f_0, f_1, f_2 \in C[\Omega, E]$ and $\Omega = [(t,u) \in J \times E : t \in J$ and $v_0 \leq u \leq w_0]$.

Definition 4.5.1. Let $v_0, w_0 \in C^1[I,E]$. Then v_0, w_0 are said to be coupled lower and upper quasi-solutions of (4.4.1) if

$$\left.\begin{aligned} v_0' &\leq f_0(t,v_0) + f_1(t,v_0) + f_2(t,w_0), v(0) \leq u_0, \\ w_0' &\geq f_0(t,w_0) + f_1(t,w_0) + f_2(t,v_0), w(0) \geq u_0. \end{aligned}\right\} \tag{4.5.2}$$

If in (4.5.2), equilities hold, then v_0, w_0 are said to be coupled quasi-solutions of (4.4.1). Clearly one can define, based on definition 4.5.1, coupled maximal and minimal quasi-solutions of (4.4.1). We also need a stronger form of coupled upper and lower quasi-solutions of (4.4.1).

Definition 4.5.2. Let $v_0, w_0 \in C^1[I,E]$ be such that $v_0(t) \leq w_0(t)$ on J. Then v_0, w_0 are said to be strongly coupled lower and upper quasi-solutions of (4.4.1) if there exists an $M > 0$ such that

$$\left.\begin{aligned} &\phi[v_0' - f_0(t,\sigma) - f_1(t,v_0) - f_2(t,w_0) + M(v_0 - \sigma)] \leq 0, \\ &\phi[w_0' - f_0(t,\sigma) - f_1(t,w_0) - f_2(t,v_0) + M(w_0 - \sigma)] \geq 0, \end{aligned}\right\} \tag{4.5.3}$$

for all σ such that $v_0(t) \leq \sigma \leq w_0(t)$, $t \in J$ and $\sigma \in K^*$.

We list for convenience the following assumptions.

(A1) For any bounded set B in E,

$$\alpha(f(I \times B)) \leq L\alpha(B);$$

(A2) $\|f(t,u_1) - f(t,u_2)\| \leq L\|u_1 - u_2\|$, (t,u_1), $(t,u_2) \in \Omega$;

A(3) $f_1(t,u)$ is nondecreasing in u and $f_2(t,u)$ is nonincreasing in u relative to K;

(A4) $f_0(t,u_1) - f_0(t,u_2) \geq -M(u_1 - u_2)$ whenever (t,u_1), $(t,u_2) \in \Omega$ and $u_2 \leq u_1$;

(A5) for any $\phi \in K^*$, $t \in J$,

$$\phi(f_0(t,u_1) - f_0(t,u_2)) = 0 \quad \text{if} \quad \phi(u_1 - u_2) = 0.$$

If f satisfies (4.5.1), (A3), (A4), and (A5), we shall say that f is a mixed quasimonotone map.

For any $\eta_1, \eta_2 \in C[J,E]$ such that $\eta_1, \eta_2 \in [v_0, w_0]$, consider the linear IVP

$$u' + Mu = \sigma(t), \quad u(0) = u_0, \tag{4.5.4}$$

where $\sigma(t) = f_0(t,\eta_1(t)) + f_1(t,\eta_1(t)) + f_2(t,\eta_2(t)) + M\eta_1(t)$.

For any $\eta_1,\eta_2 \in C[J,E]$ such that $\eta_1,\eta_2 \in [v_0,w_0]$, we define the mapping $A : [v_0,w_0] \to C[J,E]$ by

$$A[\eta_1,\eta_2] = u \tag{4.5.5}$$

where $u = u(t)$ is the unique solution of (4.5.4) on J. Concerning the map A, we prove the following lemma.

Lemma 4.5.1. Let $v_0,w_0 \in C^1[J,E]$ with $v_0(t) \leq w_0(t)$ on J. Suppose that either (a) (4.5.2), (A1), (A3), and (A4) hold, or (b) (4.5.3), (A1), and (A3) hold. Then A maps the conical segment $[v_0,w_0]$ into itself. In case of (a), A possesses further a mixed monotone property on $[v_0,w_0]$, that is, $A[\eta_1,\eta_2] \leq A[\eta_2,\eta_1]$ whenever $\eta_1 \leq \eta_2$ and $\eta_1,\eta_2 \in [v_0,w_0]$.

Proof. Let $\eta_1,\eta_2 \in C[J,E]$ be such that $\eta_1,\eta_2 \in [v_0,w_0]$ and let $u = A[\eta_1,\eta_2]$ where $u = u(t)$ is the unique solution of (4.5.4) on J. We set $p = \phi(u - v_0)$ for $\phi \in K^*$ and note that $p(0) \geq 0$. If (a) holds, we have

$$\begin{aligned} p' &\geq \phi[f_0(t,\eta_1) + f_1(t,\eta_1) + f_2(t,\eta_2) - M(u - \eta_1) - f_0(t,v_0) \\ &\qquad - f_1(t,v_0) - f_2(t,w_0)] \\ &\geq \phi[-M(\eta_1 - v_0) + f_1(t,v_0) + f_2(t,w_0) - M(u - \eta_1) - f_1(t,v_0) \\ &\qquad - f_2(t,w_0)] = -Mp. \end{aligned}$$

If, on the other hand, (b) is satisfied, one gets for all σ such that $v_0(t) \leq \sigma \leq w_0(t)$, $p' \geq \phi[f_0(t,\eta_1) + f_1(t,\eta_1) + f_2(t,\eta_2) - M(u - \eta_1) - f_0(t,\sigma) - f_1(t,v_0) - f_2(t,w_0) - M(v_0 - \sigma)]$, choosing $\sigma = \eta_1$, we arrive at

$$p' \geq \phi[f_0(t,\eta_1) + f_1(t,v_0) + f_2(t,w_0) - M(u - \eta_1) - f_0(t,\eta_1)$$

$$- f_1(t,v_0) - f_2(t,v_0) + M(v_0 - \eta_1)] = -Mp.$$

Consequently, in both cases, we get $p(t) \geq p(0)e^{-Mt} \geq 0$ on J. Since $\phi \in K^*$ is arbitrary, this proves that $v_0 \leq u$ on J. A similar argument yields that $u \leq w_0$ on J. We thus have $A[v_0,w_0] \subset [v_0,w_0]$. It remains to prove that A is mixed monotone. Let $\eta_1 \leq \eta_2$, $u_1 = A[\eta_1,\eta_2]$ and $u_2 = A[\eta_2,\eta_1]$. We set $p = \phi(u_2 - u_1)$ and note $p(0) = 0$ where $\phi \in K^*$. Then, we obtain using (A3)

$$p' = \phi[f_0(t,\eta_2) + f_1(t,\eta_2) + f_2(t,\eta_1) - M(u_2 - \eta_2) - f_0(t,\eta_1)$$

$$- f_1(t,\eta_1) - f_2(t,\eta_2) + M(u_1 - \eta_1)]$$

$$\geq \phi[-M(\eta_2 - \eta_1) - M(u_2 - \eta_2) + M(u_1 - \eta_1)] = -Mp.$$

This shows as before $p(t) \geq 0$ on J, proving $A[\eta_1,\eta_2] \leq A[\eta_2,\eta_1]$. The proof of the lemma is complete. □

In view of Lemma 4.5.1, we can define the sequences $\{v_n\},\{w_n\}$ as follows:

$$v_{n+1} = A[v_n,w_n], w_{n+1} = A[w_n,v_n] \text{ and } v_n,w_n \in [v_0,w_0].$$

The following results relative to the sequences $\{v_n\},\{w_n\}$ hold. Their proofs are similar to the corresponding results in Sections 4.4.

<u>Lemma 4.5.2</u>. Let K be normal and let the assumptions of Lemma 4.5.1 be satisfied. Then the sequences $\{v_n\},\{w_n\}$ are uniformly bounded, equicontinuous, and relative compact on J.

<u>Lemma 4.5.3</u>. Let K be normal and let the assumption (b) of Lemma 4.5.1 be satisfied. If (A2) holds, then $m(t) = 0$ on J where either

$$m(t) = \limsup_{n\to\infty} \|v_n(t) - v_{n-1}(t)\|$$

or

$$m(t) = \limsup_{n\to\infty} \|w_n(t) - w_{n-1}(t)\|.$$

In case (a), one can easily see that $\{v_n\},\{w_n\}$ are monotone sequences such that

$$v_0 \le v_1 \le \ldots \le v_n \le w_n \le \ldots \le w_1 \le w_0 \quad \text{on} \quad J.$$

Hence, in this case, the whole sequences $\{v_n\},\{w_n\}$ converge uniformly and monotonically to continuous functions ρ and r, that is,

$$\lim_{n\to\infty} v_n(t) = \rho(t), \quad \lim_{n\to\infty} w_n(t) = r(t) \quad \text{on} \quad J.$$

If case (b) holds, Lemmas 4.5.2 and 4.5.3 and (A2) permit us to conclude that the whole sequences $\{v_n\},\{w_n\}$ converge uniformly to $\rho(t)$, $r(t)$, respectively, on J. It is then easy to show from (4.5.4) that $\rho(t)$ and $r(t)$ are coupled solutions of (4.4.1) on J in each case.

We can now state our main results.

Theorem 4.5.1. Let K be normal and let $v_0,w_0 \in C^1[J,E]$ with $v_0(t)$ $w_0(t)$ on J. If the assumptions (a) (4.5.2), (A1), (A3), and (A4) hold, then there exist sequences $\{v_n\},\{w_n\}$ which converge uniformly and monotonically to coupled minimal and maximal quasi-solutions (ρ,r) of (4.4.1) on J; that is, (u_1,u_2) is any coupled quasi-solutions such that $v_0 \le u_1$, $u_2 \le w_0$ on J, then

$$v_0 \le v_1 \le \ldots \le v_n \le \rho \le u_1,\ u_2 \le r \le w_n \le \ldots \le w_1 \le w_0 \quad \text{on} \quad J,$$

provided $v_0(0) \le u_0 \le w_0(0)$. If, in addition to (a), the condition (A2) is satisfied, then $\rho(t) = r(t) = u(t)$ and $v_0 \le u \le w_0$ on J. If, on the other hand, the assumptions (b), (4.5.3), (A1), (A2), and (A3) hold, then there exists a unique solution $u(t)$ of (4.4.1) on J such that $v_0(t) \leq u(t) \leq w_0(t)$ on J provided $v_0(0) \leq u_0 \leq w_0(0)$.

Proof. In view of foregoing lemmas, because of (A2) it follows that in case (b), $u(t) = \rho(t) = r(t)$ and $v_0(t) \leq u(t) \leq w_0(t)$ on J. In case (a), we have to show that (ρ,r) are coupled minimal and maximal quasi-solutions of (4.4.1) on J. To this end, let (u_1,u_2) be any coupled quasi-solution of (4.4.1) such that u_1, $u_2 \in [v_0,w_0]$ on J. Assume that for some integer $k > 0$, we have $v_{k-1} \leq u_1$, $u_2 \leq w_{k-1}$ on J. Then setting $p = \phi(u_1 - v_k)$ and noting that $p(0) = 0$, we get

$$p' = \phi[f_0(t,u_1) + f_1(t,u_1) + f_2(t,u_2) - f_0(t,v_{k-1}) - f_1(t,v_{k-1})$$
$$- f_2(t,w_{k-1}) + M(v_k - v_{k-1})]$$
$$\geq -Mp.$$

This implies that $p(t) \geq 0$ and proves $v_k \leq u_1$ on J. A similar argument shows that $v_k \leq u_1$, $u_2 \leq w_k$ on J. Since $v_0 \leq u_1$, $u_2 \leq w_0$ on J by assumption, it follows by induction that $\rho < u_1$, $u_2 \leq r$ on J, proving that (ρ,r) are coupled minimal and maximal quasi-solutions of (4.4.1) on J. If (A2) holds, in addition, one easily sets $r(t) = \rho(t) = u(t)$ and $v_0 \leq u \leq w_0$ on J. This completes the proof of the theorem. □

It is not difficult to see that Theorem 4.5.1 contains several important special cases.

4.6 *Cone-valued Lyapunov Functions and Stability Theory.*

Let $E = R^n$ and $K \subset R^n$ be a solid cone. We consider the IVP

$$u' = g(t,u), \quad u(t_0) = u_0, \tag{4.6.1}$$

where $g \in C[R_+ \times K, R^n]$.

We observe that the quasimonotonicity of $g(t,u)$ relative to a cone P need not imply the quasimonotonicity of $g(t,u)$ relative to the cone Q if $P \subset Q$. However, the order relations relative to P imply the same order relations relative to Q whenever $P \subset Q$. From this observation results the following corollary of Theorem 4.1.3 which is very useful in applications.

Corollary 4.6.1. Let P,Q be two solid cones in R^n such that $P \subset Q$. Assume that

(a) $g \in C[R_+ \times K, R^n]$, $g(t,u)$ is quasimonotone nondecreasing in u relative to P for each $t \in R_+$, and $[t_0,\infty)$ is the largest interval of existence for the maximal solution $r(t,t_0,u_0)$ of (4.6.1) relative to P;

(b) $m \in C[R_+,R^n]$ and $D_-m(t) \underset{P}{\leq} g(t,m(t))$, $t \geq t_0$. Then $m(t_0) \underset{P}{\leq} u_0$ implies $m(t) \underset{Q}{\leq} r(t,t_0,u_0)$, $t \geq t_0$.

Consider the differential system

$$x' = f(t,x), x(t_0) = x_0, \tag{4.6.2}$$

where $f \in C[R_+ \times S(\rho), R^n]$ and $S(\rho) = \{x \in R^N : \|x\| < \rho\}$. Let K be a cone in R^n, $n \leq N$ and let $V \in C[R_+ \times S(\rho), K]$. Define for $(t,x) \in R_+ \times S(\rho)$,

$$D^+V(t,x) \equiv \limsup_{h \to 0^+} \frac{1}{h} [V(t + h, x + hf(t,x)) - V(t,x)].$$

The following comparison theorem plays a prominent role whenever we employ cone-valued Lyapunov functions.

Theorem 4.6.1. Assume that

(i) $V \in C[R_+ \times S(\rho), K]$, $V(t,x)$ satisfies a local Lipschitz condition in x relative to K and for $(t,x) \in R_+ \times S(\rho)$,

$$D^+V(t,x) \underset{K}{\leq} g(t,V(t,x)); \tag{4.6.3}$$

(ii) $g \in C[R_+ \times K, R^n]$ and $g(t,u)$ is quasimonotone in u with respect to K for each $t \in R_+$.

If $r(t,t_0,u_0)$ is the maximal solution of (4.6.1) relative to K and $x(t,t_0,x_0)$ is any solution of (4.6.2) such that $V(t_0,x_0) \underset{K}{\leq} u_0$, then, on the common interval of existence, we have

$$V(t,x(t,t_0,x_0)) \underset{K}{\leq} r(t,t_0,u_0). \tag{4.6.4}$$

<u>Proof</u>. Let $x(t) = x(t,t_0,x_0)$ be any solution of (4.6.2) such that $V(t_0,x_0) \underset{K}{\leq} u_0$. Set $m(t) = V(t,x(t))$. Then, for small $h > 0$, we have, using the fact that $V(t,x)$ is locally Lipschitzian in x relative to K,

$$\begin{aligned} m(t+h) - m(t) \underset{K}{\leq} L\|x(t+h) - x(t) - hf(t,x(t))\| + V(t+h,x(t) \\ + hf(t,x(t))) - V(t,x(t)). \end{aligned}$$

From this follows the differential inequality

$$D^+m(t) \underset{K}{\leq} g(t,m(t)) \tag{4.6.5}$$

in view of the condition (4.6.3). Now, applying Corollary 4.6.1 with $P = Q$, we get (4.6.4).

The next theorem, a variant of Theorem 4.6.1, is more flexible in applications. We merely state the result.

<u>Theorem 4.6.2</u>. Let P and Q be two solid cones in R^n such that $P \subset Q$. Suppose that

(i) $V \in C[R_+ \times S(\rho), Q]$, $V(t,x)$ satisfies a local Lipschitz condition in x relative to P and

$$D^+V(t,x) \underset{P}{\leq} g(t,V(t,x)), (t,x) \in R_+ \times S(\rho);$$

(ii) $g \in C[R_+ \times Q, R^n]$ and $g(t,u)$ is quasimonotone in u with respect to P for each $t \in R_+$.

If $r(t,t_0,u_0)$ is the maximal solution of (4.6.1) relative to P and $x(t,t_0,x_0)$ is any solution of (4.6.2) such that $V(t_0,x_0) \underset{P}{\leq} u_0$, then

$$V(t,x(t,t_0,x_0)) \underset{Q}{\leq} r(t,t_0,u_0), \tag{4.6.6}$$

on the common interval of existence of $r(t,t_0,u_0)$ and $x(t,t_0,x_0)$. If, in particular, $Q = R^n_+$, then the relation (4.6.6) implies the componentwise inequalities

$$V(t,x(t,t_0,x_0)) \leq r(t,t_0,u_0).$$

Remark 4.6.1. The comparison Theorem 4.6.1 in the special case $K = R^n_+$ has been fruitfully employed in connection with the use of vector Lyapunov functions (see Lakshmikantham and Leela [1].) In this situation, the comparison system $g(t,u)$ is required to satisfy the quasimonotone nondecreasing property in u for each $t \in R_+$, that is, for each $i = 1,2,\ldots,n$ the function $g_i(t,u_1,\ldots,u_i,\ldots,u_n)$ is nondecreasing in u_j, $i \neq j$. This comparison technique, which is called *the method of vector Lyapunov functions*, is known to be a very flexible mechanism for studying the qualitative properties of differential systems as well as for effectively discussing interconnected dynamical systems or large-scale competitive systems. However, the requirement that $g(t,u)$ be quasimonotone is too restrictive for many applications. If $g(t,u) = Au$ where A is a $n \times n$ matrix, this condition implies that the off-diagonal elements of A must be nonnegative. Since this property of A is not a necessary condition for a matrix to be stable, the limitation of this technique is clear. A possible way to get around this difficulty is to choose an appropriate cone other than R^n_+ with which to work in a given situation.

Having established Comparison Theorems 4.6.1 and 4.6.2, we can now discuss various qualitative properties including stability results by the method of cone-valued Lyapunov functions. We shall state a typical stability result.

Theorem 4.6.3. Let the assumptions (i) and (ii) of Theorem 4.6.1 hold. Suppose further that

(a) $f(t,0) \equiv 0$ and $g(t,0) \equiv 0$;

(b) for some $\phi_0 \in K_0^*$ and $(t,x) \in R_+ \times S(\rho)$,

$$b(\|x\|) \leq (\phi_0, V(t,x)) \leq a(t,\|x\|), \tag{4.6.7}$$

where $a \in C[R_+ \times [0,\rho), R_+]$, $b \in C[[0,\rho), R_+]$, $a(t,0) \equiv 0$, $b(0) = 0$ and $a(t,u)$, $b(u)$ are increasing in u;

(c) the trivial solution $u \equiv 0$ of (4.6.1) is ϕ_0-equistable, that is, given $\varepsilon > 0$ and $t_0 \in R_+$, there exists a $\delta = \delta(t_0,\varepsilon) > 0$ such that

$$(\phi_0, u_0) < \delta \quad \text{implies} \quad (\phi_0, r(t,t_0,u_0)) < \varepsilon, \quad t \geq t_0.$$

Then the trivial solution $x \equiv 0$ of (4.6.2) is equistable.

The following version of Theorem 4.6.3 is in a more flexible setting so as to be useful in applications.

Theorem 4.6.4. Let the assumptions of Theorem 4.6.2 and assumption (a) of Theorem 4.6.3 hold. Assume further that the condition (4.6.7) is satisfied for some $\phi_0 \in Q_0^*$ and that the trivial solution $u \equiv 0$ of (4.6.1) is ϕ_0-equistable, with $\phi_0 \in Q_0^*$. Then, the trivial solution $x \equiv 0$ of (4.6.2) is equistable.

The proofs of the above theorems follow the standard pattern of Lyapunov theory with appropriate modifications. They depend crucially on Theorems 4.6.1 and 4.6.2 repsectively.

A few remarks are now in order.

Remark 4.6.2. If $K = R^n_+$, $\phi_0 = (1,1,\ldots,1)$. Theorem 4.6.3 is the well-known result in the method of vector Lyapunov functions. In this case, condition (4.6.7) reduces to the familiar assumption

$$b(\|x\|) \leq \sum_{i=1}^{n} V_i(t,x) \leq a(t,\|x\|).$$

Remark 4.6.3. One could also use other measures in place of $(\phi_0, V(t,x))$. For example, let $\Phi \in C[K,R_+]$, $\Phi(u)$ is nondecreasing in u relative to K, i.e., $u \underset{K}{\leq} v$ implies $\Phi(u) \leq \Phi(v)$. The condition (4.6.7) now becomes

$$b(\|x\|) \leq \Phi(V(t,x)) \leq a(t,\|x\|).$$

Correspondingly, assumption (c) of Theorem 4.6.3 has to be modified in terms of Φ.

Remark 4.6.4. Let $P \subset Q = R^n_+$ in Theorem 4.6.4. Then, the unpleasant fact concerning the quasimonotonicity of $g(t,u)$, mentioned in Remark 4.6.1 can be removed. This, of course, means that we have to choose an appropriate cone P which necessarily depends on the nature of $g(t,u)$.

Let us demonstrate this by means of an example.

Example 4.6.1.

Consider the system

$$\begin{aligned} u_1' &= a_{11}u_1 + a_{12}u_2 \equiv g_1(t,u_1,u_2), u_1(t_0) = u_{10}, \\ u_2' &= a_{21}u_1 + a_{22}u_2 \equiv g_2(t,u_1,u_2), u_2(t_0) = u_{20}. \end{aligned} \tag{4.6.8}$$

Let $Q = R^2_+$. Suppose that we do not demand that a_{21} and a_{12} be nonnegative. Then the function $g(t,u)$ violates the quasimonotone nondecreasing condition in $u = (u_1,u_2)$ relative to Q. Hence, the differential inequalities

$$D^+V_1(t,x) \le g(t,V_1(t,x),V_2(t,x)),$$
$$D^+V_2(t,x) \le g_2(t,V_1(t,x),V_2(t,x)) \tag{4.6.9}$$

do not yield the componentwise estimates of $V(t,x(t))$ in terms of the solution of (4.6.8).

Suppose now that there exist two numbers α,β such that $0 < \beta < \alpha$ and

$$\alpha^2 a_{21} + \alpha a_{22} \ge \alpha a_{11} + a_{12}\ , \tag{4.6.10}$$

$$\beta^2 a_{21} + \beta a_{22} \ge \beta a_{11} + a_{12}\ . \tag{4.6.11}$$

These conditions can hold with no restriction of nonnegativity of a_{21} and a_{12}. We shall now choose the cone $P \subset Q = R_+$ defined by

$$P = \{u \in R_+^2 : \beta u_2 \le u_1 \le \alpha u_2\}\ .$$

This cone has two boundaries $\alpha u_2 = u_1$, and $\beta u_2 = u_1$. On the boundary $\alpha u_2 = u_1$, we take $\phi = (-1/\alpha,1)$ so that $(-1/\alpha,1)(u_1,u_1/\alpha) = 0$ and

$$(-1/\alpha,1)\ (a_{11}u_1 + a_{12}u_1/\alpha, a_{21}u_1 + a_{22}u_1/\alpha) \ge 0, \quad \text{for all} \quad u \ne 0.$$

This reduces to the condition (4.6.10). Similarly, we can obtain (4.6.11). Thus, if the inequalities (4.6.9) are relative to P, we obtain the componentwise estimates on V as

$$V_i(t,x(t)) \le r_i(t,t_0,V(t_0,x_0)), \tag{4.6.12}$$

by Theorem 4.6.2. We note that the estimate (4.6.12) is precisely the one we would have obtained if $a_{12}, a_{21} \ge 0$, by the standard method of vector Lyapunov functions. Needless to say that since a_{12}, a_{21} need not be nonnegative in our case, that method breaks down and we cannot get (4.6.12).

4.7 *Notes and Comments.*

The results of Sections 4.1 and 4.2 are taken from Lakshmikantham and Leela [1], where as the contents of Section 4.3 are adopted from Ladde, Lakshmikantham, and Vatsala [1]. The counter example in Section 4.3 is due to Volkman [1] and Theorem 4.3.2 is a classical result of Muller. Theorem 4.4.1 is due to Du and Lakshmikantham [1] where as Theorem 4.4.2 is a result of Lakshmikantham and Leela [3]. For the existence of extremal solutions and comparison results when the cone is not assumed solid see Deimling and Lakshmikantham [1,2]. The results of Section 4.5 are taken from Lakshmikantham, Leela, and Vatsala [1] while the contents of Section 4.6 are adopted from Lakshmikantham and Leela [2]. For coupled fixed points of mixed monotone oeprators see Guo and Lakshmikantham [1]. For related results see Sun and Sun [1]. For allied results of multivalued differential equations see Aubin and Cellina [1] and Deimling [1,2]. For comparison results of reaction-diffusion equations see Lakshmikantham [1]. For the use of cone-valued metric or normed spaces see Eisenfeld and Lakshmikantham [1,2,3], Nemeth [1], and Rus [1].

APPENDIX

A.1 *Separation Theorems of Convex Sets.*

Let E be a topological vector space and E^* be its dual space, i.e., the space of all linear continuous real-valued functionals on E. For any $f \in E^*$, $f \neq \theta$ (θ denotes the zero element of E^*) and any real number c, the set $H = \{x \in E \mid f(x) = c\}$ is called a closed hyperplane. The level sets $F_c = \{x \in E \mid f(x) \leq c\}$ and $F^c = \{x \in E \mid f(x) \geq c\}$ are called the closed half-spaces determined by H. The sets $G_c = \{x \in E \mid f(x) < c\}$ and $G^c = \{x \in E \mid f(x) > c\}$ are open half-spaces. Two nonempty subsets A and B of E are said to be separated by the closed hyperplane H if either $A \subset F_c$ and $B \subset F^c$ or $B \subset F_c$ and $A \subset F^c$. They are said to be strictly separated by H if either $A \subset G_c$ and $B \subset G^c$ or $B \subset G_c$ and $A \subset G^c$.

First Separation Theorem. Let A and B be two nonempty convex subsets of a topological vector space E such that $\overset{o}{A} \neq \phi$ and $\overset{o}{A} \cap B = \phi$, where $\overset{o}{A}$ denotes the interior of A. Then there exists a closed hyperplane H which separates A and B. Moreover, H strictly separates $\overset{o}{A}$ and B.

Second Separation Theorem. Let A and B be two nonempty, disjoint convex subsets of a locally convex topological vector space E such that A is closed and B is compact. Then there exists a closed hyperplane H strictly separating A and B.

For the proofs of these theorems and related results see Schaefer [1], Dunford and Schwartz [1], and Pascali and Sburlan [1].

A.2 *Zorn's Lemma.*

Let P be a set of elements. Suppose there is a binary relation defined between certain pairs of element a,b of P, expressed symbolically by $a \leq b$, with the properties:

(a) if $a \leq b$ and $b \leq c$, then $a \leq c$;

(b) $a \leq a$ for any $a \in P$;

(c) if $a \leq b$ and $b \leq a$, then $a = b$.

Then P is said to be partially ordered by the relation.

If P is partially ordered and if, moreover, for every pair a,b of P either $a \leq b$ or $b \leq a$, then P is said to be completely ordered.

A subset of a partially ordered set P is itself partially ordered by the relation which partially orders P. Also, a subset of P may turn out to be completely ordered by this relation.

If P is partially ordered set and S is a subset of P, an element $c \in P$ is called an upper bound of S if $a \leq c$ for every $a \in S$. An element $c \in P$ is said to be maximal if $a \in P$ and $c \leq a$ together imply $c = a$.

Zorn's Lemma. Let P be a nonempty partially ordered set with the property that every completely ordered subset of P has an upper bound in P. Then P contains at least one maximal element.

It is well known that Zorn's Lemma is equivalent to the axiom of choice. It is immaterial whether we choose to regard Zorn's Lemma as an axiom of set theory or as a theorem derived from the axiom of choice.

For Zorn's Lemma mentioned in detail see Taylor [1] and Dunford and Schwartz [1].

A.3 *Leray-Schauder Degree.*

(a) Brouer Degree.

Let Ω be a nonempty bounded open set in R^n and $f : \overline{\Omega} \to R^n$ be a C^2 mapping, where $\overline{\Omega}$ denotes the closure of Ω. Let $p \in R^n \setminus f(\partial\Omega)$, where $\partial\Omega$ denotes the boundary of Ω. Then $\tau = d(p, f(\partial\Omega)) = \inf_{\tau \in \partial\Omega} \|f(x) - p\| > 0$. Construct a continuous function $\phi : [0,+\infty) \to R^1$, which satisfies the following conditions:

(i) there exist $0 < \alpha < \beta \leq \tau$ such that $\Phi(r) = 0, \quad r \notin (\alpha,\beta)$;

(ii) $\int_{R_n} \Phi(\|z\|)dz = 1$.

Now we define the Brouwer degree $\deg(f,\Omega,p)$ by

$$\deg(f,\Omega,p) = \int_{\overline{\Omega}} \Phi(\|f(x) - p\|)J_f(x)dx, \tag{A.3.1}$$

where $J_f(x)$ denotes the Jacobian of f at x, i.e., $J_f(x) = \det f'(x)$.

We can prove that $\deg(f,\Omega,p)$ is an integer and is independent of the choice of Φ.

By the method of approximation we can extend the definition of Brouwer degree from C^2 mappings to C mappings. Let $f : \overline{\Omega} \to R^n$ be a continuous mapping and $p \in R^n \setminus f(\partial\Omega)$, then

$$\tau = d(p,f(\partial\Omega)) = \inf_{x \in \partial\Omega} \|f(x) - p\| > 0. \tag{A.3.2}$$

Choose a C^2 mapping $g : \overline{\Omega} \to R^n$ such that

$$\max_{x \in \overline{\Omega}} \|f(x) - g(x)\| < \tau. \tag{A.3.3}$$

It is easy to know $p \in R^n \setminus g(\partial\Omega)$, hence $\deg(g,\Omega,p)$ is well defined. We now define the <u>Brouwer degree</u> of f on Ω with respect to p, i.e., $\deg(f,\Omega,p)$, to be $\deg(g,\Omega,p)$:

$$\deg(f,\Omega,p) = \deg(g,\Omega,p). \tag{A.3.4}$$

It is not difficult to prove that $\deg(g,\Omega,p)$ is independent of the choice of g. When $\Omega = \phi$, we define $\deg(f,\Omega,p) = 0$.

<u>Theorem A.3.1</u>. The Brouwer degree defined by (A.3.4) has the following properties:

(i) Normality: $\deg(I,\Omega,p) = 1$ for any $P \in \Omega$, where I denotes the identical mapping, i.e., $Ix = x, \ \forall\, x \in \overline{\Omega}$.

(ii) Additivity: $\deg(f,\Omega,p) = \deg(f,\Omega_1,p) + \deg(f,\Omega_2,p)$ whenever Ω_1, and Ω_2 are disjoint open subsets of Ω such that $p \notin f(\overline{\Omega}\backslash(\Omega_1 \cup \Omega_2))$.

(iii) Homotopy invariance: $\deg(h(t,\cdot),\Omega,p)$ is independent of $t (0 \leq t \leq 1)$ whenever $h : [0,1]\times \Omega \to R^n$ is continuous and $p \notin h(t,\partial\Omega)$ for any $t \in [0,1]$.

(iv) Change of base: $\deg(f,\Omega,p) = \deg(f - p,\Omega,\theta)$, where θ denotes the zero element of R^n.

Moreover, let

$$M = \{(f,\Omega,p) \mid \Omega \subset R^n \text{ open bounded, } f : \overline{\Omega} \to R^n \text{ continuous, } p \in R^n\backslash f(\partial\Omega)\}$$

and Z be the set of integers. Then there exists exactly one function $d : M \to Z$, the Brouwer degree, satisfying (i)-(iv). In other words, the Brouwer degree is unique.

By method of homeomorphism we can extend the definition of Brouwer degree from Euclidean space to finite-dimensional Banach space. Let E_n be an n-dimensional real Banach space and Ω be a bounded open set of E_n. Suppose that operator $f : \overline{\Omega} \to E_n$ is continuous and $p \in E_n \backslash f(\partial\Omega)$. We choose a basis of E_n : $\{e_1,e_2,\ldots e_n\}$, and define a mapping h by $hx = y$, where $x = \sum_{i=1}^{n} \alpha_i e_i \in E_n$ and $y = (\alpha_1,\alpha_2,\ldots,\alpha_n) \in R^n$. Evidently, h is a linear homeomorphic mapping from E_n onto R^n. Now, we define

$$\deg(f,\Omega,p) = \deg(hfh^{-1},h(\Omega),h(p)), \tag{A.3.5}$$

the right hand is a Brouwer degree in R^n. It is easy to prove that $\deg(f,\Omega,p)$ is independent of the choice of basis $\{e_i\}$ and Theorem A.3.1 is still true for $\deg(f,\Omega,p)$ defined by (A.3.5).

(b) Leray-Schauder Degree.

Let Ω be a bounded open set in real Banach space E and $F : \overline{\Omega} \to E$ be completely continuous. Let $f = I - F$ and $p \in E \setminus f(\partial\Omega)$. Then (A.3.2) holds. Choose a bounded and continuous operator $F_n : \overline{\Omega} \to E^{(n)}$ such that

$$\|F(x) - F_n(x)\| < \tau, \quad x \in \overline{\Omega},$$

where $E^{(n)}$ is a finite-dimensional subspace of E and $p \in E^{(n)}$. Let $\Omega_n = E^{(n)} \cap \Omega$ and $f_n(x) = x - F_n(x)$. It is easy to show that $f_n : \overline{\Omega}_n \to E^{(n)}$ is continuous and $p \in E^{(n)} \setminus f_n(\partial\Omega_n)$ and hence, the Brouwer degree $\deg(f_n,\Omega_n,p)$ in $E^{(n)}$ is well defined. Now, we define the Leray-Schauder degree $\deg(f,\Omega,p)$ to be $\deg(f_n,\Omega_n,p)$, i.e.,

$$\deg(f,\Omega,p) = \deg(f_n,\Omega_n,p). \tag{A.3.6}$$

We can prove that $\deg(f,\Omega,p)$ is independent of the choice of F_n. Similar to Theorem A.3.1, we have

Theorem A.3.2. The Leray-Schauder degree defined by (A.3.6) has the following properties:

(i) Normality: $\deg(I,\Omega,p) = 1$ for any $p \in \Omega$.

(ii) Additivity: $\deg(f,\Omega,p) = \deg(f,\Omega_1,p) + \deg(f,\Omega_2,p)$ whenever Ω_1 and Ω_2 are disjoint open subsets of Ω such that $p \notin f(\overline{\Omega} \setminus (\Omega_1 \cup \Omega_2))$.

(iii) Homotopy invariance: $\deg(h(t,\cdot),\Omega,p)$ is independent of $t (0 \le t \le 1)$ whenever $h(t,x) = x - H(t,x)$ and $H : [0,1] \times \overline{\Omega} \to E$ is completely continuous and $p \notin h(t,\partial\Omega)$ for any $t \in [0,1]$.

(iv) Change of base: $\deg(f,\Omega,p) = \deg(f - p,\Omega,\theta)$, where θ denotes the zero element of E.

Moreover, let

$$M = \{(f,\Omega,p) \mid \Omega \subset E \text{ openbounded, } f = I - F \text{ and } F : \overline{\Omega} \to E \text{ completely continuous, } p \in E \setminus f(\partial\Omega)\}$$

and Z be the set of integers. Then there exists exactly one function $d : M \to Z$, the Leray-Schauder degree, satisfying (i)-(iv). In other words, the Leray Schauder degree is unique.

Theorem A.3.3. Besides (i)-(iv), the Leray-Schauder degree has the following properties:

(v) Solution property: if $\deg(f,\Omega,p) \neq 0$, then the equation $f(x) = p$ has at least one solution in Ω.

(vi) Excision property: $\deg(f,\Omega,p) = \deg(f,\Omega_0,p)$ whenever Ω_0 is an open subset of Ω such that $p \notin f(\overline{\Omega} \setminus \Omega_0)$.

(vii) Boundary value property: $\deg(f,\Omega,p) = \deg(g,\Omega,p)$ whenever $f = I - F$, $g = I - G$ and $F(x) = G(x)$ for any $x \in \partial\Omega$.

(viii) Connected component property: $\deg(f,\Omega,\cdot)$ is constant on every connected component of $E \setminus f(\partial\Omega)$.

(ix) Property of omitting a direction: $\deg(f,\Omega,p) = 0$ if there exists $\theta \neq y_0 \in E$ such that $f(x) \neq p + ty_0$ for any $x \in \partial\Omega$ and $t \geq 0$.

(c) Topological Degrees of Strict-Set-Contraction Fields and condensing Fields.

By means of the Leray-Schauder degree we can extend the conception of topological degree to strict-set-contraction fields.

Let Ω be a bounded open set in real Banach space E and $F: \overline{\Omega} \to E$ is a k-set-contraction, where $0 \leq k < 1$, i.e., $\gamma(F(S)) \leq k\gamma(S)$ for any $S \subset \overline{\Omega}$, where $\gamma(S)$ denotes the measure of noncompactness of S. Let $f = I - F$ and f is called to be a k-set-contraction field.

First we assume $\theta \in E \setminus f(\partial\Omega)$ and define $\deg(f,\Omega,\theta)$ as follows. Let $D_1 = \overline{co}\, F(\overline{\Omega})$, and define, by induction,

$$D_n = \overline{co}\, F(D_{n-1} \cap \overline{\Omega}) \quad (n = 2,3,4,\ldots) .$$

If $D_n \cap \overline{\Omega} = \phi$ for some n, we define the topological degree

$$\deg(f,\Omega,\theta) = 0. \tag{A.3.7}$$

Now, suppose $D_n \cap \overline{\Omega} \neq \phi$ for any n. Then, we can prove the measure of noncompactness $\alpha(D_n) \to 0$, and so the set $D = \bigcap_{n=1}^{\infty} D_n$ is nonempty, convex, and compact. Moreover, $D \cap \overline{\Omega}$ is nonempty and compact, and $F(D \cap \overline{\Omega}) \subset D$. Hence, by the extension theorem, there exists a completely continuous operator $F_1 : \overline{\Omega} \to D$ such that $F_1(x) = F(x)$ for any $x \in D \cap \overline{\Omega}$. Let $f_1 = I - F_1$ and it is easy to show that $\theta \notin f_1(\partial\Omega)$, and therefore the Leray-Schauder degree $\deg(f_1,\Omega,\theta)$ is well defined. We define

$$\deg(f,\Omega,\theta) = \deg(f_1,\Omega,\theta). \tag{A.3.8}$$

It is not difficult to prove that $\deg(f,\Omega,\theta)$ is independent of the choice of F_1.

Next, we assume $p \in E \setminus f(\partial\Omega)$. In this case, we define

$$\deg(f,\Omega,p) = \deg(f - p,\Omega,\theta). \tag{A.3.9}$$

The right hand of (A.3.9) is well defined since $f(x) - p \neq \theta$ for any $x \in \partial\Omega$.

We can prove that Theorems A.3.2 and A.3.3 are still true for topological degree of k-set-contraction fields $(0 \leq k < 1)$ defined by (A.3.8) and (A.3.9). Notice, in this case, the homotopy invariance property takes the following form: $\deg(h(t,\cdot),\Omega,p)$ is independent of t $(0 \leq t \leq 1)$ whenever $h(t,x) = x - H(t,x)$ and $H(t,x)$ satisfied

(a) $H : [0,1] \times \overline{\Omega} \to E$ is continuous and the continuity of $H(t,x)$ in t is uniform with respect to $x \in \overline{\Omega}$, (b) $H(t,\cdot) : \overline{\Omega} \to E$ is a k-set-contraction, where $0 \leq k < 1$ and k is independent of $t \in [0,1]$ and (c) $p \notin h(t,\partial\Omega)$ for any $t \in [0,1]$.

By means of the topological degree of k-set-contraction fields $(0 \leq k < 1)$, we can extend the conception of topological degree to condensing fields.

Let Ω be a bounded open set in real Banach space E and $f = I - F$, where $F : \overline{\Omega} \to E$ is condensing, i.e., $\gamma(F(S)) < \gamma(S)$ for any $S \subset \overline{\Omega}$ with $\gamma(S) > 0$, where $\gamma(S)$ denotes the measure of noncompactness of S. Let $p \in E \setminus f(\partial\Omega)$. Choose a k-set-contraction $G : \overline{\Omega} \to E$ $(0 \leq k < 1)$ such that

$$\|F(x) - G(x)\| < \tau, \quad \forall\, x \in \overline{\Omega},$$

where

$$\tau = d(p,f(\partial\Omega)) = \inf_{x \in \partial\Omega} \|f(x) - p\| > 0 .$$

Such a G exists, for example, we can choose $G = kF$, where $0 \leq k < 1$ and $1 - k$ is sufficiently small. Let $g = I - G$. Then g is a k-set-contraction field and it is easy to see $p \notin g(\partial\Omega)$. Hence $\deg(g,\Omega,p)$ is well defined. Now, we define

$$\deg(f,\Omega,p) = \deg(g,\Omega,p) . \tag{A.3.10}$$

It is easy to show that $\deg(f,\Omega,p)$ is independent of the choice of G. Also, we can prove that Theorem A.3.2 and A.3.3 are still true for topological degree of condensing fields defined by (A.3.10). In this case, the homotopy invariance property takes the following form: $\deg(h(t,\cdot),\Omega,p)$

is independent of t $(0 \leq t \leq 1)$ whenever $h(t,x) = x - H(t,x)$ and $H(t,x)$ satisfies (a) $H : [0,1] \times \overline{\Omega} \to E$ is continuous and the continuity of $H(t,x)$ in t is uniform with respect to $x \in \overline{\Omega}$, (b) $H(t,\cdot) : \overline{\Omega} \to E$ is condensing for any $t \in [0,1]$ and (c) $p \notin h(t,\partial\Omega)$ for any $t \in [0,1]$.

For the theory of Leray-Schauder degree and related results, see Cronin [1], Schwartz [1], Nirenberg [1], Krasnosel'skii and Zabreiko [1], Guo [1], Deimling [1], and Lloyd [1].

A.4 *Properties of Nemitskii Operators.*

Let G be a Lebesque measurable set in R^n such that $0 < \text{mes}\, G \leq \infty$ A function $f(x,u)$, where $x \in G$, $-\infty < u < +\infty$, is said to satisfy the Caratheodory condition if it satisfies the following two conditions:

(a) $f(x,u)$ is a continuous function with respect to u for almost all $x \in G$;

(b) $f(x,u)$ is a measurable function with respect to x for any $u \in (-\infty,+\infty)$.

In the following, we always assume that $f(x,u)$ satisfies the Caratheodory condition.

The operator defined by

$$f\phi(x) = f(x,\phi(x)) \tag{A.4.1}$$

is called the Nemitskii operator.

Theorem A.4.1. If f maps $L^{p_1}(G)$ $(p_1 \geq 1)$ into $L^{p_2}(G)$ $(p_2 \geq 1)$, then f is continuous.

Theorem A.4.2. If f maps $L^{p_1}(G)$ $(p_1 \geq 1)$ into $L^{p_2}(G)$ $(p_2 \geq 1)$, then f is bounded.

Theorem A.4.3. f maps $L^{p_1}(G)$ $(p_1 \geq 1)$ into $L^{p_2}(G)$ $(p_2 \geq 1)$ if and only if $f(x,u)$ satisfies

$$|f(x,u)| \leq a(x) + b|u|^{p_1/p_2} \quad (x \in G, \quad -\infty < u < +\infty), \tag{A.4.2}$$

where $a(x) \geq 0$, $a(x) \in L^{p_2}(G)$ and $b > 0$.

For the proofs of these theorems and related results, see Krasnosel'skii [2] and Guo [1], [4].

A.5 *Extension Theorems.*

The following two extension theorems are useful.

Theorem A.5.1. Let E_1 and E_2 be two real Banach spaces and D be a closed subset of E_1. Suppose that operator $A : D \to E_2$ is completely continuous. Then there exists a completely continuous operator $\tilde{A} : E_1 \to E_2$ such that $\tilde{A}x = Ax$ for any $x \in D$ and $\tilde{A}(E_1) \subset \overline{co}\, A(D)$, where $\overline{co}\, A(D)$ denotes the closed convex hull of $A(D)$.

Theorem A.5.2. Let E_1 be a metric space and D be a closed subset of E_1. Let E_2 be a locally convex topological vector space. Suppose that map $A : D \to E_2$ is continuous. Then there exists a continuous map $\tilde{A} : E_1 \to E_2$ such that $\tilde{A}x = Ax$ for any $x \in D$ and $\tilde{A}(E_1) \subset \overline{co}\, A(D)$.

For the proofs of these theorems and related results, see Dugundji [1] and Guo [1].

REFERENCES

Amann, H.

[1] On the number of solutions of nonlinear equations in ordered Banach spaces, J. Funct. Anal., 11(1972), 346-384.

[2] Fixed point equations and nonlinear eigenvalue problems in ordered Banach spaces, SIAM Review, 18(1976), 620-709.

Ambrosetti, A., and Rabinowitz, P.H. [1] Dual variational methods in critical point theory and applications, J. Funct. Anal., 14(1973), 349-381.

Aubin, J. P., and Cellina, A. [1] Differential Inclusions, Springer-Verlag, New York, 1984.

Bai, Jindong, [1] Eigenvectors of nonlinear integral operators of Hammerstein type, Kexue Tongbao, 29(1984), 704.

Berger, M. S., [1] Nonlinear problems with exactly three solutions, Indiana Univ. Math. J., 28(1979), 689-698.

Bernfeld, S. R., and Lakshmikantham, V. [1] An Introduction to Nonlinear Boundary Value Problems, Academic Press, New York, 1974.

Brezis, H., and Browder, F. E. [1] Nonlinear integral equations and systems of Hammerstein type, Adv. Math., 18(1975), 115-147.

Bushell, P. J.

[1] Hilbert's metric and positive contraction mappings in a Banach space, Arch. Rational Mech. Anal., 52(1973), 330-338.

[2] On a class of Volterra and Fredholm nonlinear integral equations, Math. Proc. Camb. Phil. Soc., 79(1976), 329-335.

Cac, N. P., and Gupta, J. A. [1] Fixed point theorems for mappings in ordered Banach spaces, J. Math. Anal. Appl., 71(1979(, 547-557.

Cahlon, B., and Eskin, M. [1] Existence theorems for an integral equation of the Chandrasekhar H-equation with perturbation, _J. Math. Anal. Appl._, 83(1981), 159-171.

Cohen, D. [1] Multiple stable solutions of nonlinear boundary value problems arising in chemical reactor theory, _SIAM J. Appl. Math._, _20_(1971), 1-13.

Cronin, J.

[1] Fixed points and topological degree in nonlinear analysis, _Amer. Math. Soc._, 1964.

[2] Eigenvalues of some nonlinear operators, _J. Math. Anal._, _38_(1972), 659-667.

Deimling, K.

[1] _Nonlinear Functional Analysis_, Springer-Verlag, New York, 1985.

[2] Multivalued maps and multivalued differential equations, University of Texas at Arlington Technical Report #233, April, 1985.

[3] Positive fixed points of weakly inward maps, _Nonlinear Anal._, 12(1988), 223-226.

Deimling, K., and Hu, Shou-Chuan, [1] Fixed points of weakly inward maps in conical shells, _Nonlinear Anal._, 12(1988), 227-230.

Deimling, K., and Lakshmikantham, V.

[1] On the existence of extremal solutions of differential equations in Banach spaces, _Nonlinear Anal._, _3_(1979), 563-568.

[2] Existence and comparison results for differential equations in Banach spaces, _Nonlinear Anal._, 3(1979), 569-600.

Du, S. W., and Lakshmikantham, V. [1] Monotone iterative technique for differential equations in a Banach space, _J. Math. Anal. Appl._, _87_(1982), 454-459.

Dugundji, J. [1] An extension of Tietze's Theorem, Pacific J. Math., 1(1951), 353-367.

Dunford, N., and Schwartz, J. T. [1] Linear Operators, Part I: General Theory, Interscience Publishers, New York, 1958.

Eisenfeld, J., and Lakshmikantham, V.

[1] Comparison principle and nonlinear contractions in abstract spaces, J.M.A.A., 49(1975), 504-511.

[2] Fixed point theorems through abstract cones, J.M.A.A., 52(1975), 25-35.

[3] Fixed point theorems on closed sets through abstract cones, Appl. Math. Comp., 3(1977), 155-167.

[4] Differential inequalities and flow invariance via linear functionals, J.M.A.A., 69(1979), 106-116.

Gilbarg, D., and Trudinger, N. S. [1] Elliptic Partial Differential Equations of Second Order, Springer-Verlag, New York, 1977.

Crandall, M. G., and Rabinowitz, P. H. [1] Multiple solutions of a nonlinear integral equation, Arch. Rational Mech. Anal., 37(1970), 262-267.

Guo, Dajun.

[1] Nonlinear Functional Analysis (in Chinese), Shandong Sci. Tech. Publishing House, China, 1985.

[2] Positive solutions of nonlinear operator equations and its applications to nonlinear integral equations (in Chinese), Adv. Math., 13(1984), 294-310.

[3] A fixed point theorem of decreasing operators and its applications (in Chinese), Kexue Tongbao, 29(1984), 189.

[4] Properties of the Nemitskii operators and its applications (in Chinese), Adv. Math., 6(1963), 70-91.

[5] The number of positive solutions of Hammerstein nonlinear integral equations (in Chinese), Acta Math. Sinica, 22(1979), 584-595.

[6] Fixed points and eigenvectors of come classes of concave and convex operators (in Chinese), Kexue Tongbao, 30(1985), 1132-1135.

[7] Eigenvalues and eigenvectors of nonlinear operators, Chinese Ann. Math., 2(eng. Issue), 1981), 65-80.

[8] Some fixed point theorems and applications, Nonlinear Anal., 10(1986), 1293-1302.

[9] Multiple positive solutions of nonlinear integral equations and applications, Applic. Anal., 23(1986), 77-84.

[10] Some fixed point theorems on cone maps, Kexue Tongbao, 29(1984), 575-578.

[11] The number of nontrivial solutions of Hammerstein integral equations and their applications, Kexue Tongbao, 27(1982), 694-98.

[12] A new fixed point theorem (in Chinese), Acta Math. Sinica, 24(1981), 444-450.

[13] On the solution of a nonlinear integral equation in neutron transport theory (in Chinese), Acta Math. Sinica, 22(1979), 231-236.

[14] The solution of a nonlinear integral equation in nuclear physics (in Chinese), Kexue Tongbao, 23(1978), 27-31.

[15] Positive eigenvectors of decreasing operators and applications, Northeastern Math. J., 1(1985), 101-109.

[16] The number of nontrivial solutions of Hammerstein nonlinear integral equations, Chinese Ann. Math., 7B(2)(1986), 191-204.

[17] Solvability of Hammerstein integral equations with applications (in Chinese), Acta Math. Sinica, 16(1966), 137-149.

[18] Positive solutions of Hammerstein integral equations of polynomial type with applications (in Chinese), Chinese Ann. Math., 4A(5)(1983), 645-656.

[19] Eigenvalues and eigenfunctions of Hammerstein integral equations (in Chinese), Acta Math. Sinica, 25(1982), 419-426.

Guo, Dajun, and Lakshmikantham, V. [1] Coupled fixed points of nonlinear operators with applications, Nonlinear Anal., 11(1987), 623-632.

Guo, Dajun, and Qingyong, Zhang. [1] On the uniqueness of solution of a nonlinear integral equation in nuclear physics (in Chinese), Kexue Tongbao, 24(1979), 678-681.

Hardy, G. H., Littlewood, J.E., and Polya, G. [1] Inequalities, Cambridge University Press, 1934.

Hively, G. A. [1] On a class of nonlinear integral equations arising in transport theory, SIAM J. Math. Anal., 9(1978), 787-792.

Huang, Chunchao. [1] A generalization of the Guo Dajun theorem, Kexue Tongbao, 29(1984), 1341.

Krasnosel'skii, M. A.

[1] Positive solutions of operator equations, P. Noordhoff, Groningen, The Netherlands, 1964.

[2] Topological methods in the theory of nonlinear integral equations, Pergamon Press, Oxford, 1964.

Krasnosel'skii, M. A., and Zabreiko, P. P. [1] Geometrical methods of nonlinear analysis, Springer-Verlag, New York, 1984.

Krein, M. G., and Rutman, M. A. [1] Linear operators leaving invariant a cone in a Banach space, Amer. Math. Soc. Trans., 10(1962), 199-325.

Ladde, G. S., Lakshmikantham, V., and Vatsala, A. S. [1] Monotone iterative techniques for nonlinear differential equations, Pitman Publishing, Belmont, Ca., 1985.

Lakshmikantham, V. [1] Comparison theorems for reaction-diffusion equations, Proc. SAFA Conference, Bari, Italy, Gius. Laterza and Figli, S.P.A. Bari (1979), 121-156.

Lakshmikantham, V., and Leela, S.

[1] Nonlinear differential equations in abstract spaces, Pergamon Oxford, 1981.

[2] Cone valued Lyapunov functions, Nonlinear Anal., 1(1977), 215-222.

[3] On the method of upper and lower solutions in abstract cones, Ann. Polon. Math., 42(1983), 159-164.

[4] Differential and Integral Inequalities, Vol. I-II, Academic Press, New York, 1969.

[5] Method of quasi-upper and lower solutions in abstract cones, Nonlinear Anal., 6(1982), 833-838.

Leatsch, T. [1] Existence and bounds for multiple solutions of nonlinear equations, SIAM J. Appl. Math., 18(1970), 389-400.

Leggett, R. W., and Williams, L. R.

[1] Multiple positive fixed points of nonlinear operators on ordered Banach space, Indiana Univ. Math. J., 28(1979), 673-688.

[2] A fixed point theorem with application to an infectious disease model, J. Math. Anal. Appl., 76(1980), 91-97.

[3] An extension of Jentzch's Theorem to nonlinear Hammerstein operators, J. Math. Anal. Appl., 60(1977), 248-254.

Lloyd, N. G. [1] Degree Theory, Cambridge University Press, 1978.

Massabo, I., and Stuart, C. A. [1] Positive eigenvectors of k-set-contractions, Nonlinear Anal., 3(1979), 35-44.

Miller, R. F. [1] On a nonlinear integral equation occurring in diffraction theory, Proc. Camb. Phil. Soc., 62(1966), 249-261.

Nemeth, A. B. [1] Normal Cone valued matrices and nonconvex vector minimization principle, Seminar of functional analysis and numerical methods, Cluj-Napoea, Rumania(1983), 117-156.

Nirenberg, L.

[1] Topics in nonlinear functional analysis, Lecture Notes, Courant Institute, New York, 1974.

[2] Variational and topological methods in nonlinear problems, Bull. Amer. Math. Soc., 4(1981), 267-300.

Nussbaum, R. D. [1] The fixed point index for local condensing maps, Ann. Mat. Pura Appl., 89(1971), 217-258.

Pascali, D., and Sburlan, S. [1] Nonlinear mappings of monotone type, Editura Academici, Romania, 1978.

Pazy, A., and Rabinowitz, P.H.

[1] A nonlinear integral equation with applications to neutron transport theory, Arch. Rational Mech. Anal., 32(1969), 226-246.

[2] Corrigendum: A nonlinear integral equation with applications to neutron transport theory, Arch. Rational Mech. Anal., 35(1969), 409-410.

Peressini, A. L. [1] Ordered topological vector spaces, Harper and Row, New York, 1968.

Potter, A. J. B. [1] Applications of Hilbert's projective metric to certain classes of non-homogeneous operators, Quart. J. Math. Oxford, (2), 28(1977), 93-99.

Qin, Chenglin [1] Positive fixed points of quasi-homogeneous mappings, Acta Math. Sinica, 27(1984), 792-794.

Rus, I. A. [1] Principles and applications of fixed point theory (Romanian), Dacia, Cluj-Napoca, 1979.

Schaefer, H. H. [1] Topological vector spaces, Springer-Verlag, New York, 1971.

Stuart, C. A. [1] Positive solutions of a nonlinear integral equation, Math. Ann., 192(1971), 119-124.

Sun, Jingxian.

[1] On the equivalence of normal cones and fully regular cones in reflexive Banach spaces (in Chinese), Kexue Tongbao, 28(1984), 382.

[2] A generalization of Guo's theorem and applications, unpublished.

Sun, Jingxian, and Sun, Yong. [1] Some fixed point theorems of increasing operators, Applic. Anal., 23(1986), 23-27.

Swick, K. E. [1] A model of single species population growth, SIAM J. Math. Anal., 7(1976), 565-576.

Taylor, A. E. [1] Introduction to functional analysis, John Wiley and Sons, New York, 1958.

Thieme, H. R. [1] On a class of Hammerstein integral equations, Manuscripta Math., 29(1979), 49-84.

Vainberg, M. M. [1] Variational methods for the study of nonlinear operators, Holden-Day, San Francisco, 1964.

Volkmann, P. [1] Ausdehnung eines Satzes von Max Muller auf unendliche Systeme von gewohnlichen Differentialgleichunger, Funkcial. Ekvac. 21(1978), 81-96.

Wan, Weixun. [1] Contraction conditions of mapping and fixed point theorems of Banach type (in Chinese), Acta Math. Sinica, 27(1984), 35-52.

Williams, L. R., and Leggett, R. W.

[1] Multiple fixed point theorems for problems in chemical reactor theory, *J. Math. Anal. Appl.*, *69*(1979), 180-193.

[2] Nonzero solutions of nonlinear integral equations modeling infectious disease, *SIAM J. Math. Anal.*, *13*(1982), 112-121.

Yu, Qingyu, and Browder, F. E. [1] Boundary conditions for condensing mappings, *Nonlinear Anal.*, *8*(1984), 209-219.

Zaanen, A. C. [1] *Linear Analysis*, North-Holland, Amsterdam, 1953.

NOTES AND REPORTS IN MATHEMATICS IN SCIENCE AND ENGINEERING

Edited by William F. Ames, Georgia Institute of Technology

Vol. 1 John S. Gero, editor, *Design Optimization*

Vol. 2 Michael F. Barnsley and Stephen G. Demko, *Chaotic Dynamics and Fractals*

Vol. 3 Sigeru Mizohata, *On the Cauchy Problem*

Vol. 4 Heinz W. Engl and C.W. Groetsch, editors, *Inverse and Ill-Posed Problems*

Vol. 5 Dajun Guo and V. Lakshmikantham, *Nonlinear Problems in Abstract Cones*

U.C.W. ABERYSTWYTH
LIBRARY